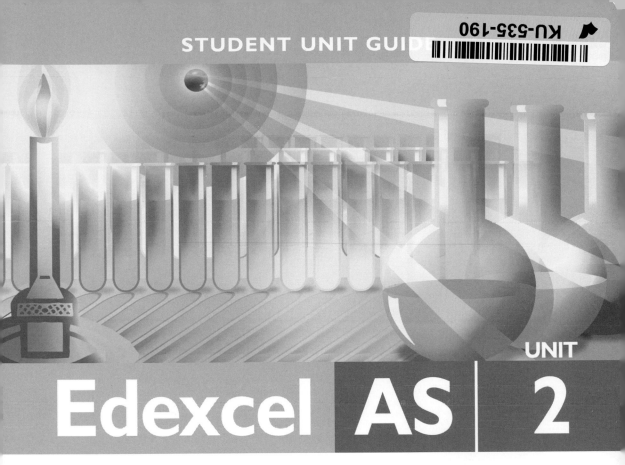

Edexcel AS | UNIT 2

Chemistry

Application of Core Principles of Chemistry

Rod Beavon

Philip Allan Updates, an imprint of Hodder Education, an Hachette UK Company, Market Place, Deddington, Oxfordshire OX15 0SE

Orders

Bookpoint Ltd, 130 Milton Park, Abingdon, Oxfordshire, OX14 4SB
tel: 01235 827827
fax: 01235 400401
e-mail: education@bookpoint.co.uk
Lines are open 9.00 a.m.–5.00 p.m., Monday to Saturday, with a 24-hour message answering service. You can also order through the Philip Allan Updates website: www.philipallan.co.uk

© Philip Allan Updates 2008

ISBN 978-0-340-94818-7

First printed 2008
Impression number   5   4
Year   2013   2012   2011

This guide has been written specifically to support students preparing for the Edexcel AS Chemistry Unit 2 examination. The content has been neither approved nor endorsed by Edexcel and remains the sole responsibility of the author.

Printed by MPG Books, Bodmin

Hachette UK's policy is to use papers that are natural, renewable and recyclable products and made from wood grown in sustainable forests. The logging and manufacturing processes are expected to conform to the environmental regulations of the country of origin.

Contents

Introduction

■ ■ ■

Content Guidance

■ ■ ■

Questions and Answers

Introduction

About this guide

This unit guide is the second of a series covering the Edexcel specification for AS and A2 chemistry. It offers advice for the effective study of **Unit 2: Application of Core Principles of Chemistry**. Its aim is to help you *understand* the chemistry. It is not intended as a shopping list, enabling you to cram for an examination. The guide has three sections:

- **Introduction** — this provides guidance on study and revision, together with advice on approaches and techniques to ensure that you answer examination questions in the best way that you can.
- **Content Guidance** — this section is not intended to be a textbook. It offers guidelines on the main features of the content of Unit 2, together with particular advice on making study more productive.
- **Questions and Answers** — this section shows you the sort of questions you can expect in the unit test. Answers are provided with examiner comments. Careful consideration of these answers and comments will improve your answers but, more importantly, will improve your understanding of the chemistry involved.

The effective understanding of chemistry requires time. No one suggests that chemistry is an easy subject, but if you find it difficult you can overcome your problems by the proper investment of your time.

To understand the chemistry, you have to make links between the various topics. The subject is coherent and is not a collection of discrete modules. Once you have spent time thinking about chemistry, working with it and solving chemical problems, you will become aware of these links. Spending time this way will make you fluent with the ideas. Once you have that fluency, and practise the good techniques described in this book, the examination will look after itself. Don't be an examination automaton — be a chemist.

The specification

The specification describes the chemistry that can be examined in the unit tests and the format of those tests. This is not necessarily the same as what teachers might choose to teach or what you might choose to learn.

The purpose of this book is to help you with Unit Test 2, but don't forget that what you are doing is learning *chemistry*. The specification can be obtained from Edexcel, either as a printed document or from the web at **www.edexcel.org.uk**.

The unit test

The unit test consists of a structured question paper of duration 1 hour 15 minutes, worth 80 marks. This unit counts for 40% of AS or 20% of Advanced GCE marks. Section A contains objective test (multiple-choice) questions and section B has a mixture of short-answer and extended-answer questions. Section C consists of questions based on a contemporary context and assesses quality of written communication.

Assessment objectives

Unit Test 2 has three assessment objectives (AO). AO1 is 'knowledge and understanding of science and of How Science Works'. This makes up 40% of the test. You should be able to:

- recognise, recall and show understanding of scientific knowledge
- select, organise and present information clearly and logically, using specialist vocabulary where appropriate

AO2 is 'application of knowledge and understanding of science and of How Science Works'. This makes up 50% of Unit Test 2. You should be able to:

- analyse and evaluate scientific knowledge and processes
- apply scientific knowledge and processes to unfamiliar processes including those related to issues
- assess the validity, reliability and credibility of scientific information

AO3 is 'How Science Works'. This makes up 10% of Unit Test 2. You should be able to:

- demonstrate and describe ethical, safe and skilful practical techniques and processes, selecting appropriate qualitative and quantitative methods
- make, record and communicate reliable and valid observations and measurements with appropriate precision and accuracy
- analyse, interpret, explain and evaluate the methodology, results and impact of your own and others' experimental and investigative activities in a variety of ways

How Science Works

How Science Works (HSW) is not new. It involves using theories and models, posing questions that can be answered scientifically, carrying out practical investigations and being able to use scientific terminology properly. All this has featured in previous specifications, but in this new one the role of HSW in assessment and its proportion in an examination paper are specified. Except for the new material mentioned later, it's unlikely that you will notice any difference between HSW questions and questions on similar topics from past papers.

An example of HSW is in the material on shapes of molecules. This uses the idea of minimising repulsions between the electron pairs about a given atom. A carbon atom having four electron pairs gives a molecular shape that puts these pairs as far apart as possible, so if all the bonds are single bonds the molecule is tetrahedral, like methane or carbon tetrachloride. This idea can be extended to other molecules to predict their shapes, which is HSW because it is a general theory that enables predictions to be made.

Material that is new in this specification is:
- the assessment of risk in experiments and understanding the difference between risk and hazard
- considering the way in which science is communicated and how the quality of the work is judged by other scientists ('peer review')
- 'green' chemistry, particularly the efforts to develop processes that are less hazardous and make more efficient use of resources both in terms of the proportion of the atoms that finish up in the product ('atom economy') and in terms of lower energy consumption.

Command terms

The following command terms are used in the specification and in unit test questions. You must distinguish between them carefully.
- **Recall** — a simple remembering of facts learned, without any explanation or justification of these facts.
- **Understand** — be able to explain the relationship between facts and underlying chemical principles (understanding enables you to use facts in new situations).
- **Predict** — say what you think will happen on the basis of learned principles.
- **Define** — give a simple definition, without any explanation.
- **Determine** — find out.
- **Show** — relate one set of facts to another set.
- **Interpret** — take data or other types of information and use them to construct chemical theories or principles.
- **Describe** — state the characteristics of a particular material or thing.
- **Explain** — use chemical theories or principles to say why a particular property of a substance or series of substances is as it is.

Learning to learn

Learning is not instinctive — you have to develop suitable techniques to make your use of time effective. In particular, chemistry has peculiar difficulties that need to be understood if your studies are to be effective from the start.

Planning

Efficient people do not achieve what they do by approaching life haphazardly. They plan — so that if they are working, they mean to be working, and if they are watching television, they have planned to do so. Planning is essential. You must know what you have to do each day and set aside time to do it. Furthermore, to devote time to study means you may have to give something up that you are already doing. There is no way that you can generate extra hours in the day.

Be realistic in your planning. You cannot work all the time and you must build in time for recreation and family responsibilities.

Targets

When devising your plan, have a target for each study period. This might be a particular section of the specification, or it might be rearranging of information from text into pictures, or drawing a flowchart relating all the reactions of group 2 and group 7. Whatever it is, be determined to master your target material before you leave it.

Reading chemistry textbooks

A page of chemistry may contain a range of material that differs widely in difficulty. Therefore, the speed at which the various parts of a page can be read is variable. You should read with pencil and paper to hand and jot things down as you go — for example, equations, diagrams and questions to be followed up. If you do not write down the questions, you will forget them; if you do not master detail, you will never become fluent in chemistry.

Text

This is the easiest part to read. Little advice is needed here.

Chemical equations

Equations are used because they are quantitative, concise and internationally understood. Take time over them, copy them and check that they balance. Most of all, try to visualise what is happening as the reaction proceeds. If you can't, make a note to ask someone who can or — even better — ask your teacher to *show* you the reaction if possible. Equations describe real processes; they are not abstract algebraic constructs.

Graphs

Graphs give a lot of information and they must be understood in detail rather than as a general impression. Take time over them. Note what the axes are, what the units are, the shape of the graph and what the shape means in chemical terms.

Tables

These are a means of displaying a lot of information. You need to be aware of the table headings and the units of numerical entries. Take time over them. What trends can be seen? How do these relate to chemical properties? Sometimes it can be useful to convert tables of data into graphs.

Diagrams

Diagrams of apparatus should be drawn in section. When you see them, copy them and ask yourself why the apparatus has the features it has. What is the difference between a distillation and a reflux apparatus, for example? When you do practical work, examine each piece of the apparatus closely so that you know both its form and its function.

Mathematical equations

In chemistry, mathematical equations describe the real, physical world. If you do not understand what an equation means, ask someone who does.

Calculations

Do not take calculations on trust — work through them. First, make certain that you understand the problem and then that you understand each step in the solution. Make clear the units of the physical quantities used and make sure you understand the underlying chemistry. If you have problems, ask.

Always make a note of problems and questions that you need to ask your teacher. Learning is not a contest or a trial. Nobody has ever learnt anything without effort or without running into difficulties from time to time — not even your teachers.

Notes

Most people have notes of some sort. Notes can take many forms: they might be permanent or temporary; they might be lists, diagrams or flowcharts. You have to develop your own styles. For example, notes that are largely words can often be recast into charts or pictures and this is useful for imprinting the material. The more you rework the material, the clearer it will become.

Whatever form your notes take, they must be organised. Notes that are not indexed or filed properly are useless, as are notes written at enormous length and those written so cryptically that they are unintelligible a month later.

Writing

In chemistry, extended writing is often not required. However, you need to be able to write concisely and accurately. This requires you to marshal your thoughts properly and needs to be practised during your day-to-day learning.

Have your ideas assembled in your head before you start to write. You might imagine them as a list of bullet points. Before you write, have an idea of how you are going to link these points together and also how your answer will end. The space available for an answer is a poor guide to the amount that you have to write — handwriting sizes differ hugely, as does the ability to write succinctly. Filling the space does not necessarily mean you have answered the question. The mark allocation suggests the number of points to be made, not the amount of writing needed.

Re-reading

When you have completed your work, you must re-read it critically. This is remarkably difficult, because you tend to read what you intended to write rather than what you actually did write. Nevertheless, time spent on the evaluation of your own work is time well spent. You should be able to eliminate at least the majority of silly errors

— but you need to practise this in your day-to-day work and not do it for the first time in an examination.

Approaching the unit test

The unit test is designed to allow you to show the examiner what you know. Answering questions successfully is not only a matter of knowing the chemistry but also a matter of technique.

Revision

Start your revision in plenty of time. Make a list of the things that you need to do, emphasising the things that you find most difficult, and draw up a detailed revision plan. Work back from the examination date, ideally leaving an entire week free from fresh revision before that date. Be realistic in your revision plan and then add 25% to the timings because everything takes longer than you think.

When revising:
- make a note of difficulties and ask your teacher about them, otherwise you will forget to ask
- practise answering questions from past papers, but remember that they have been written to a different specification
- revise ideas, rather than forms of words — you are after *understanding*
- remember that scholarship requires time to be spent on the work

When you use the example questions in this book, make a determined effort to answer them before looking up the answers and comments. Remember that the answers here are not intended as model answers to be learnt parrot-fashion. They are designed to illuminate chemical ideas and understanding.

The exam

The exam paper
- *Read the whole paper through before you start to write.* Even though there is no choice of questions, knowing what is around the corner helps the brain to do some useful subconscious processing.
- *Read the question.* Questions usually change from one examination to the next. A question that looks the same, at a cursory glance, as one that you have seen before usually has significant differences when read carefully. Needless to say, candidates do not receive credit for writing answers to their own questions.
- Be aware of the number of marks available for a question. This is an excellent pointer to the number of points that you need to make.
- Do not repeat the question in your answer. There is a danger that you fill up the space available and think that you have answered the question, when you might have ignored some or maybe all of the real points.

- The name of a 'rule' is not an explanation for a chemical phenomenon. Thus, in equilibrium a popular answer to a question on the effect of a change of pressure on an equilibrium system is 'because of Le Chatelier's principle...'. That is simply a name for a rule — it does not explain anything.

Multiple-choice questions

Answers to multiple-choice questions are machine-marked. Multiple-choice questions need to be read carefully and it is important not to jump to a conclusion about the answer too quickly. You need to be aware that one of the options might be a 'distracter'. An example is in a question having a numerical answer of, say, -600 kJ mol^{-1}. A likely distracter would be $+600$ kJ mol^{-1}.

Some questions require you to think on paper. You do not have to work out answers to multiple-choice questions in your head. Space is provided on the question paper for rough working. It will not be marked, so do not write anything that matters in this space, because the examiner will not see it.

For each of the questions there are four suggested answers, A, B, C and D. You select the *best* answer by putting a cross in the box beside the letter of your choice. If you change your mind you should put a line through the box and then indicate your alternative choice. Multiple choices score zero. This format is not used in the questions in this book.

Unit 2 has 20 multiple-choice questions that should take you no more than 20 minutes.

Marking

Online marking

It is important that you have some understanding of how examinations are marked, because to some extent it affects how you answer them. Your examination technique partly concerns chemistry and partly must be geared to how the examinations are dealt with physically. You have to pay attention to the layout of what you write. Because all your scripts are scanned and marked online, there are certain things you must do to ensure that all your work is seen and marked. This is covered below.

As the examiner reads your answer, decisions have to be made — is this answer worth the mark or not? Those who think that these decisions are always easy 'because science is right or wrong' have misunderstood the nature of marking and the nature of science.

Your job is to give the *clearest possible answer* to the question asked, in such a way that your chemical understanding is made obvious to the examiner. In particular, you must not expect the examiner to guess what is in your head; you can be judged only by what you write.

Not all marking is the same

The marking of homework is not the same as the marking of examinations. Teachers marking their pupils' work are engaging in formative assessment and their marking is geared towards helping students to improve their understanding of chemistry. It will include comments and suggestions for improving under-standing, which are far more important than any mark that might be obtained, as is any discussion resulting from the work. An examination is a summative assess-ment; candidates have no opportunity to improve, so the mark is everything. That is why questions that are designed to improve the understanding of chemistry during the course are not of the same style as questions used to test that under-standing at the end of a course.

Please do not regard examination questions and answers as chemical education in their own right; they are an attempt to see if you have acquired that education through the influence of your teachers and your own reading. An examination is a means to an end, not an end in itself. Education, as distinct from training, is designed to make you (in this case) into a competent chemist rather than one who can simply regurgi-tate 'model answers' with no underlying understanding. The truth is that if you do the necessary work throughout the course the examination content should look after itself. What you need is the proper answering techniques.

Because examination answers cannot be discussed, you must make your answers as clear as possible. Do not expect examiners to guess what is in your head. This is one reason why you are expected, for example, to show working in calculations. It is especially important that you *think before you write*. You will have a space on the question paper that the examiner has judged to be a reasonable amount for the answer. Because of differing handwriting sizes, false starts and crossings-out and because some candidates have a tendency to repeat the question in the answer space, the space is never exactly right for all candidates. Advice on avoiding some of the pitfalls comes later; but the best advice is that *before you begin your answer you must have a clear idea of how it will end*. You do not have time or space for subsequent editing. It is a good plan to practise putting your answer into a list of the points you wish to make and join them into coherent sentences. Or you can leave them as a list; good quality written communication can just as well be presented as a list as a piece of elegant prose.

Common pitfalls
- Do *not* write in any colour other than black (an exam board regulation).
- Do *not* write outside the space provided without saying, *within that space*, where the remainder of the answer can be found.

Edexcel exam scripts are marked online, so few examiners will handle a real, original script. The process is as follows:
- When the paper is set it is divided up into items, often, but not necessarily, a single part of a question. These items are also called clips.
- The items are set up so that they display on-screen, with check-boxes for the score and various buttons to allow the score to be submitted or for the item to be processed in some other way.
- After you have written your paper it is scanned; from that point all the handling of your paper is electronic. Your answers are tagged with an identity number.

- It is impossible for an examiner to identify a centre or a candidate from any of the information supplied.
- Examiners mark items over a period of about 3 weeks.
- Examiners are instructed on how to apply the marking scheme and are tested to make sure that they know what is required and can mark the paper fairly.
- Examiners are monitored on their performance throughout the marking period. They are prevented from marking a particular item if they do not achieve the necessary standard of accuracy; their defective marking is re-marked.
- Examiners mark items, not whole scripts. This style of marking means that an individual candidate's paper could be marked by as many as 20 different people.
- Items are allocated to examiners randomly, so generally they do not see more than one item from a given candidate.

The following list of potential pitfalls to avoid is particularly important:

- Do *not* write in any colour other than black. The scans are entirely black-and-white, so any colour used simply comes out black — unless you write in red, in which case it does not come out at all. The scanner cannot see red (or pink or orange) writing. So if, for example, you want to highlight different areas under a graph, or distinguish lines on a graph, you must use a different sort of shading rather than a different colour.
- Do *not* use small writing. Because the answer appears on a screen, the definition is slightly degraded. In particular, small numbers used for powers of 10 can be difficult to see. The original script is always available but it can take a long time to get hold of it.
- Do *not* write in pencil. Faint writing does not scan well.
- Do *not* write outside the space provided without saying, within that space, where the remainder of the answer can be found. Examiners only have access to a given item; they cannot see any other part of your script. So if you carry on your answer elsewhere but do not tell the examiner within the clip that it exists, it will not be seen. Although the examiner cannot mark the out-of-clip work, the paper will be referred to the Principal Examiner responsible for the paper.
- Do *not* use asterisks or arrows as a means of directing examiners where to look for out-of-clip items. Tell them in words. Candidates use asterisks for all sorts of things and examiners cannot be expected to guess what they mean.
- Do not write across the centre-fold of the paper from the left-hand to the right-hand page. A strip about 8 mm wide is lost when the papers are guillotined for scanning.
- Do not repeat the question in your answer. If you have a questions such as 'Define the first ionisation energy of calcium', the answer is 'The energy change per mole for the formation of unipositive ions from isolated calcium atoms in the gas phase'; or, using the equation, 'The energy change per mole for $Ca(g) \rightarrow Ca^+(g) + e^-$'. Do not start by writing 'The first ionisation energy for calcium is defined as...' because by then you will have taken up most of the space available for the answer. Examiners know what the question is — they can see it on the paper.

Content
Guidance

This section is a guide to the content of **Unit 2: Application of Core Principles of Chemistry**. The main areas of this unit are:

- Shapes of molecules and ions
- Intermediate bonding and bond polarity
- Intermolecular forces
- Redox
- The periodic table: groups 2 and 7
- Kinetics
- Chemical equilibria
- Organic chemistry
- Mechanisms
- Mass spectra and IR
- Green chemistry

For each part of the specification, you should also consult a standard textbook for more information. Chemistry is a subtle subject and you need to have a good sense of where the information you are dealing with fits into the larger chemical landscape. This only comes by extensive reading. Remember that the specification tells you only what can be examined in the unit test. You need as much practice as you can get in thinking about chemistry.

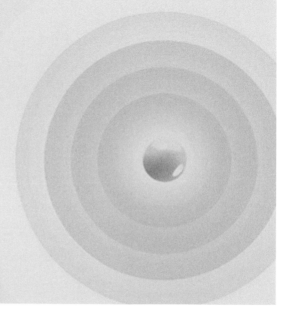

Shapes of molecules and ions

Valence-shell electron-pair repulsion theory

The shape of a molecule is determined by several factors:
- The number of electron pairs around the atom of interest, both lone pairs and bond pairs.
- The electrons arrange themselves as far apart as possible.
- Lone pair–lone pair repulsions are greater than lone pair–bond pair repulsions, which in turn are greater than bond pair–bond pair repulsions. The repulsions between electrons modify the bond angles. Thus in methane the H–C–H angle is 109.5°, in ammonia the H–N–H angle is 107° and in water the H–O–H angle is 104°.

The shapes of analogous molecules can be predicted from those given above. The shape of methane (CH_4) is the same as that of silane (SiH_4), phosphine (PH_3) has a similar shape to ammonia (NH_3) and the shape of water (H_2O) is similar to that of hydrogen sulfide (H_2S).

A double or triple bond is stereochemically equivalent to a single bond.

Species with single bonds only

The arrangement of the bonds depends on the number of electron pairs in total. The name of the shape depends on the position of the atom centres only.

Name	Formula	Bond pairs	Lone pairs	Shape	Structure
Beryllium chloride	$BeCl_2$	2	0	Linear	Cl–Be–Cl
Boron trichloride	BCl_3	3	0	Trigonal planar	
Methane	CH_4	4	0	Tetrahedral	
Hydrogen chloride	HCl	1	3	Linear	H–Cl
Ammonia	NH_3	3	1	Pyramidal	

Name	Formula	Bond pairs	Lone pairs	Shape	Structure
Ammonium ion	NH_4^+	4	0	Tetrahedral	
Water	H_2O	2	2	Bent	
Phosphorus pentachloride (gas phase only — the solid is PCl_4^+ and PCl_6^- ions)	PCl_5	5	0	Trigonal bipyramidal	
Sulfur hexafluoride	SF_6	6	0	Octahedral	

Species with double bonds

Stereochemically a double bond is equivalent to a single bond.

Name	Formula	Single bonds	Double bonds	Lone pairs	Shape	Structure
Carbon dioxide	CO_2	0	2	0	Linear	O=C=O
Sulfur dioxide	SO_2	0	2	1	Bent	
Sulfite ion	SO_3^{2-}	2	1	1	Pyramidal	

Name	Formula	Single bonds	Double bonds	Lone pairs	Shape	Structure
Carbonate ion	CO_3^{2-}	2	1	0	Trigonal planar	
Sulfate ion	SO_4^{2-}	2	2	0	Tetrahedral	
Nitrate ion	NO_3^{-}	1	2	0	Trigonal planar	

Diamond, graphite and the fullerenes

Diamond has layers of hexagonal rings that are puckered, not flat: each carbon atom is bonded to four others throughout the lattice by covalent bonds. Diamond is a poor electrical conductor because it has no delocalised electrons. It is a good thermal conductor, since the stiff lattice readily transmits vibration. It has a high melting temperature (c. 3800°C).

Graphite has layers of flat hexagons with the fourth bond delocalised along the plane of the carbons. Bonds between layers are van der Waals and therefore weak. Graphite is a stack of giant molecules. It is a good electrical conductor parallel to the planes but a poor thermal conductor. It sublimes at 3730°C.

Buckminsterfullerene (C_{60}) was discovered in the late 1990s. It is a sphere of pentagons and hexagons, exactly like a football. It is red and soluble in organic solvents.

C_{60}

Related to fullerenes are carbon nanotubes, which are tubular structures, made from carbon hexagons with diameters from one to several nanometres. Their structure is like a roll of wire netting. They can be capped at each end with a hemisphere of carbon atoms to make nanometre-sized capsules. Much research is being conducted to develop these tiny capsules so that they can deliver drugs and other treatments to tumour cells with great precision, avoiding damage to healthy tissue. Another potential application is in building electronic devices even smaller than those presently available.

Intermediate bonding and bond polarity

Electronegativity and polarity

Ionic and covalent bonds are extremes. Ionic bonding involves complete electron transfer and covalent bonding involves equal sharing of electrons in pairs. In practice, bonding is intermediate between these two forms in most compounds, with one type being predominant.

Electronegativity

The attraction a bonded atom has for electrons is called its **electronegativity**. There are several electronegativity scales, the commonest being the Pauling scale. Fluorine is the most electronegative element with a value of 4, caesium the least at 0.7. There are no units.

- Atoms with the same electronegativity bond covalently with equal sharing of electrons.
- Atoms with different electronegativity form:
  — polar covalent bonds if the difference is not too large — up to about 1.5
  — ionic bonds if the difference is more than about 2

Polarity

Cations can distort the electron clouds of anions (which are generally larger). This distortion leads to a degree of electron sharing and hence some covalence in most ionic compounds.

- Small cations with a high charge (2+ or 3+) have a high charge density. They can polarise anions and have therefore a high polarising power.
- Large anions do not hold on to their outer electrons very tightly, so they can be easily distorted. Such ions are polarisable.
- Magnesium chloride ($MgCl_2$) has a polarising cation but a not very polarisable anion. There is some covalent character in the compound, but much more in magnesium iodide (MgI_2) since the large I^- ion is highly polarisable.

Polarity in molecules

A molecule may have polar covalent bonds, but will not be polar overall if the bond polarities cancel because of the shape of the molecule. Thus boron trifluoride (BF_3) has polar bonds but no overall polarity because the trigonal planar shape means the bond polarities cancel. The same is true for the tetrahedral carbon tetrachloride (CCl_4). Ammonia (NH_3) has polar bonds and is polar overall because its pyramidal shape does not cause the individual polarities to cancel.

These molecules have polar bonds, but the polarities cancel overall because of the symmetry of the molecule.

These molecules also have polar bonds, but the individual polarities do not cancel so the molecules are polar overall.

Intermolecular forces

Types of intermolecular force

There are several types of intermolecular force. They depend to some extent on the distortion of the electron distribution of one molecule by another. At close range there are also repulsive forces in action. In decreasing order of strength, the attractive forces are:

- hydrogen bonds
- permanent dipole–permanent dipole forces
- permanent dipole–induced dipole forces
- temporary dipole–induced dipole forces

The last two forces are often referred to generically as **van der Waals** forces.

Hydrogen bonds

Hydrogen bonds are electrostatic bonds between a hydrogen atom covalently bonded to an electronegative atom (N, O or F) and another N, O or F atom, usually on a different molecule. The hydrogen is made highly δ+ and thus is attracted to *a lone pair* *of electrons* on the electronegative atom, which is δ–. The three atoms involved in the bond are collinear.

Hydrogen bond strength in hydrogen fluoride (HF) is around 150 kJ mol^{-1}, but most others are between 60 and 20 kJ mol^{-1}. Hydrogen bonding is responsible for the high boiling temperatures of NH_3, H_2O and HF compared with the other hydrides in their groups and for the high solubility of alcohols and sugars in water. The hydrogen bonds formed between the solute and the water are of similar strength to the hydrogen bonds in water itself.

Permanent dipole–permanent dipole forces
Polar molecules attract via their permanent dipoles. Polar molecules have higher boiling temperatures than non-polar molecules of similar size.

Permanent dipole–induced dipole forces
These forces arise where a polar molecule distorts the electron density on a non-polar molecule, giving rise to a temporary dipole that attracts the permanent dipole. The strength of this force depends on the polarisability of the non-polar molecule, which generally increases with increasing numbers of electrons in the molecule.

Temporary dipole–induced dipole forces
These are sometimes called **London** forces. They are the only interaction in non-polar molecules or in single atoms such as the monatomic inert or noble gases. These forces arise from a temporary dipole inducing a complementary dipole in an adjacent molecule. These dipoles are always shifting, but are induced in phase and give a net attraction. Their strength depends on the polarisability of the molecule.

Properties of materials

Intermolecular forces (including those between ions or atoms — but be clear which you are talking about) determine the physical properties of a substance such as its melting and boiling temperatures, hardness and density. The forces between particles in liquids are similar to those in solids, but because of thermal agitation they are less directional.

Change of state
In solids the particles vibrate about a mean position in the crystal lattice. Heating increases the amplitude of the vibration until a point where the interparticular forces are overcome and the lattice collapses — the solid melts. Just above the melting temperature, liquids often have considerable order to their structure; further heating causes movement over increasing distances, until the temperature is such that the vapour pressure of the liquid is the same as the external pressure. At this temperature the interparticular forces are overcome and the liquid boils.

Noble gases
The only interatomic forces are dispersion (London) forces arising from the limited movement of electrons within the atom. The smaller the atom, the less the movement

and so the weaker the attractions. This is reflected in their boiling temperatures:

Element	He	Ne	Ar	Kr	Xe	Rn
Boiling temperature/K	4	27	87	121	165	211

Hydrides of group 4

- None of the elements in group 4 is electronegative enough to give hydrogen bonding.
- The hydrides have covalent intramolecular bonds, intermolecular bonds being van der Waals (London-type) forces.
- The more electrons, the greater the attraction. The boiling temperatures are:

Hydride	Methane (CH_4)	Silane (SiH_4)	Germane (GeH_4)	Stannane (SnH_4)
Boiling temperature/K	109	161	183	221

Hydrides of groups 5–7

Nitrogen, oxygen and fluorine are electronegative enough to give hydrogen bonding in their hydrides. Therefore, for molecules of their size, these hydrides have much higher boiling temperatures than would be expected from van der Waals interactions alone. The remaining hydrides have both dipole–dipole and van der Waals interactions and in every case the latter dominate. The boiling temperature therefore increases with increasing size of the hydride.

Group 5

Hydride	NH_3	PH_3	AsH_3	SbH_3
Boiling temperature/K	240	185	218	256

Group 6

Hydride	H_2O	H_2S	H_2Se	H_2Te
Boiling temperature/K	373	212	232	269

Group 7

Hydride	HF	HCl	HBr	HI
Boiling temperature/K	293	188	206	238

Alkanes

The intermolecular forces in the non-polar alkanes are temporary dipole–induced dipole. The forces increase in strength with increasing numbers of electrons in the molecules and decrease in strength the further apart the molecules are. There is an increase in both the melting and the boiling temperatures of the straight-chain alkanes as the chain length, and hence the number of electrons, increases. The values for the melting temperature of the solids are less regular at first because of differences in the crystal packing. The values in the table below are given to three significant figures.

Alkane	Methane	Ethane	Propane	Butane	Pentane	Hexane
Melting temperature/K	91.1	89.8	83.4	135	143	178
Boiling temperature/K	109	185	231	273	309	342

Effect of branching

Since the strength of the intermolecular forces falls off rapidly with increasing distance, the melting and boiling temperature of branched-chain isomers of the alkanes are generally lower than those of the straight-chain compound (compare pentane and 2-methylbutane below). However, 2,2-dimethylpropane is a symmetrical molecule and can pack well both in the crystal lattice and in the liquid, so its values are much higher than those of 2-methylbutane.

The following table compares the melting and boiling temperatures of the three isomers of C_5H_{12}.

Alkane	Pentane	2-methylbutane	2,2-dimethylpropane
Melting temperature/K	143	113	257
Boiling temperature/K	309	301	283

Volatility of alkanes and alcohols

An alcohol is less volatile (i.e. has a higher boiling temperature) than an alkane of similar size. Alcohols are strongly hydrogen-bonded, whereas alkanes are non-polar and have only London forces between their molecules. These London forces are much weaker than hydrogen bonds.

Alkane	Ethane (CH_3CH_3)	Propane $(CH_3CH_2CH_3)$	Butane $(CH_3CH_2CH_2CH_3)$
Boiling temperature/K	185	231	273
Alcohol	Methanol (CH_3OH)	Ethanol (CH_3CH_2OH)	Propan-1-ol $(CH_3CH_2CH_2OH)$
Boiling temperature/K	338	352	371

Solute, solvent and solubility

The solubility of a substance in a given solvent depends on the enthalpy (heat energy) and entropy changes of the dissolving process. Entropy is considered at A2 in Unit 4, so solubility is discussed here only in terms of enthalpy, but you should remember that the full story is more complex.

In general a substance dissolves if the bonds formed between the solute and the solvent are of similar or greater strength than those that are broken between both the solute molecules and the solvent molecules. Thus, for a solute A and a solvent B, the A–B bonds must be similar in strength to, or stronger than, both the A–A bonds and the B–B bonds. The difference between the energy needed to break these bonds is the factor that determines the extent of the solubility.

Ionic compounds

When an ionic compound dissolves in water, the ionic lattice has to be broken up (endothermic process) and the resulting ions are then hydrated (exothermic process). For some compounds where the solubility is controlled mostly by the enthalpy changes, the following Hess's law cycle applies.

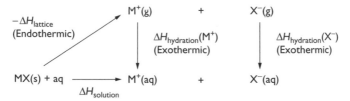

If the energy required to break up the lattice is recovered by the hydration of the ions, the compound is soluble, i.e.

$$\Delta H_{solution} = -\Delta H_{lattice} + \Delta H_{hydration}(\text{cation}) + \Delta H_{hydration}(\text{anion})$$

must be exothermic or only slightly endothermic. (Remember that this applies only where the entropy changes are less important than the enthalpy changes.)

Alcohols in water

An alcohol molecule (ROH) consists of a non-polar alkyl group and a hydroxyl group (–OH) that can hydrogen-bond with water. On dissolving, the –OH group breaks some of the water–water hydrogen bonds and replaces them with alcohol–water hydrogen bonds. The alcohol dissolves if the new alcohol–water bonds formed are strong enough to compensate energetically for the water–water bonds that are broken by the alkyl group (R) occupying space in the liquid water. Alcohols up to C_3 are miscible with water in all proportions. Butan-1-ol and its isomers are partially soluble, dissolving to some extent; alcohols above C_5 are insoluble in water. The long hydrophobic carbon chains disrupt the structure of the water too much.

Larger alcohols dissolve if they have more –OH groups. For example, glucose is a C_6 molecule, but it has 6 –OH groups so it is highly soluble in water.

Polar molecules in water

Hydrogen bonding in water is strong. To overcome this and separate the water molecules a solute has to form strong bonds with water. Polar molecules such as halogenoalkanes are not usually able to form strong enough bonds, so they are not water-soluble.

Non-aqueous solvents: 'like dissolves like'

Since the intermolecular forces in the solute are similar to those between the solvent and to those between solvent and solute in the solution, non-polar solvents such as hydrocarbons (e.g. hexane) dissolve other non-polar substances, such as waxes and fats. In contrast, hydrocarbons do not dissolve sugars, because the hydrogen bonding between the sugar molecules in the solid is too strong to be energetically compensated by any interaction between the non-polar hydrocarbon and the polar sugar molecules.

Redox

Oxidation numbers

Oxidation numbers are used to find the ratio (the stoichiometry) in which the oxidising and reducing agents react. To find the oxidation number of an element in any species, the following rules are used:
- An uncombined element has an oxidation number of zero.
- A simple ion has an oxidation number equal to its charge.
- Oxygen has oxidation number −2 (except in peroxides and superoxides).
- Hydrogen has oxidation number +1 (except in metal hydrides).

Example 1

What is the oxidation number of sulfur in sulfur trioxide (SO_3)?

Answer

Oxygen has an oxidation number of −2, which gives $S^{x+}(O^{2-})_3$, so $x = +6$. The oxidation number of S in SO_3 is +6.

Example 2

What is the oxidation state of manganese in manganate (MnO_4^-)?

Answer

A similar argument gives $[Mn^{y+}(O^{2-})_4]^-$. The oxygen contributes −8, with −1 charge overall. Thus $y = +7$. The oxidation number of Mn in MnO_4^- is +7.

The oxidation number of an atom in a compound is the charge that it would have if the compound were ionic. It is useful because changes in oxidation number indicate that an atom has been oxidised or reduced. Oxidation numbers can be used for simple atoms and ions, molecules or complex ions.

Electron transfer

The reaction of magnesium with oxygen is obviously an oxidation:

$$2Mg + O_2 \rightarrow 2MgO$$

- Magnesium is converted into magnesium ions Mg^{2+}, the oxygen to oxide ions O^{2-}.
- A logical extension of this reaction is to say that any reaction converting magnesium to its ion is an oxidation and that **oxidation is electron loss**.
- The reverse process, the acquisition of electrons, is therefore **reduction**.
- **OILRIG**: oxidation is loss, reduction is gain.

An oxidising agent removes electrons from another species, so it gains electrons in a redox process. Similarly, a reducing agent is itself oxidised and therefore loses electrons.

Changes in oxidation number

Oxidation occurs if the oxidation state of an atom rises; reduction causes the oxidation state to fall.

Example 1

Iron reacts with chlorine on heating to give iron(III) chloride.

$$2Fe \quad + \quad 3Cl_2 \quad \rightarrow \quad 2FeCl_3$$

Oxidation numbers: $\quad$ 0 $\qquad\qquad$ 0 $\qquad\qquad$ +3 −1

The oxidation number of the iron goes up, so it has been oxidised. The oxidation number of chlorine goes down, so it has been reduced.

Example 2

Disproportionation is a reaction where the oxidation number of a given species both rises and falls. In the disproportionation of chlorate(I) ions (OCl^-), the oxidation number of chlorine both rises and falls:

$$3OCl^- \quad \rightarrow \quad 2Cl^- \quad + \quad ClO_3^-$$

Oxidation number of chlorine $\quad$ +1 $\qquad\qquad$ −1 $\qquad$ +5

Two chlorine atoms in the +1 state are reduced to the −1 state and one is oxidised from +1 to +5.

Combining half-equations

Redox reactions are a combination of an oxidation and a reduction process. Each process can be written as a half-equation, which includes electrons. Two half-equations can be combined to give an overall equation. One or both of the half-equations may need to be multiplied by a small integer in order to make the number of electrons the same in both equations. This is because *the same electrons* are involved in each equation.

Example 1

The reaction of chlorine with the bromide ions in seawater is used in the commercial production of bromine.

Oxidation of bromide ions: $\quad 2Br^- \rightarrow Br_2 + 2e^-$
Reduction of chlorine: $\quad Cl_2 + 2e^- \rightarrow 2Cl^-$

Since there are two electrons in both half-reactions, they can simply be added to give the overall equation:

$$2Br^- + Cl_2 \rightarrow Br_2 + 2Cl^-$$

Example 2

Iron reacts with chlorine on heating to give iron(III) chloride.

Oxidation of iron: $Fe \rightarrow Fe^{3+} + 3e^-$
Reduction of chlorine: $Cl_2 + 2e^- \rightarrow 2Cl^-$

Since the number of electrons is not the same, the top reaction is multiplied by 2 and the bottom one by 3. The equations can then be added to give the overall reaction:

$2Fe \rightarrow 2Fe^{3+} + 6e^-$
$3Cl_2 + 6e^- \rightarrow 6Cl^-$

Overall: $2Fe + 3Cl_2 \rightarrow 2Fe^{3+} + 6Cl^-$ (or $2FeCl_3$)

The periodic table: groups 2 and 7

Group 2 (alkaline earth metals)

Ionisation energies

The ionisation energies fall with increasing atomic number (down the group). Although the nuclear charge is rising, the size of the atom is also rising, as is the amount of shielding (or repulsion) from inner-shell electrons.

Element	Be	Mg	Ca	Sr	Ba
First ionisation energy/kJ mol^{-1}	900	736	590	548	502
Second ionisation energy/kJ mol^{-1}	1760	1450	1150	1060	966

Reactions of group 2 metals

Reaction with oxygen
All the group 2 metals burn to produce the oxide. For example:

$$2Mg(s) + O_2(g) \xrightarrow{\text{Burn}} 2MgO(s)$$

- Magnesium burns vigorously with a brilliant white flame.
- Calcium burns with a brick-red flame.
- Strontium burns with a crimson flame.
- Barium burns with a pale apple-green flame, forming substantial amounts of barium peroxide (Ba_2O_2), as well as the oxide (BaO).

All the group 2 oxides are white, ionic compounds.

Reaction with chlorine

On heating with chlorine, all the group 2 metals react similarly to give white, ionic chlorides:

$$Mg(s) + Cl_2(g) \xrightarrow{\text{Heat}} MgCl_2(s)$$

Reaction with water

The vigour of the reaction decreases down the group.

- Magnesium reacts slowly with cold water, but rapidly with steam.

$$Mg(s) + H_2O(g) \xrightarrow{\text{Heat}} MgO(s) + H_2(g)$$

- Calcium reacts quite quickly with cold water to give a milky suspension of calcium hydroxide, some of which dissolves.

$$Ca(s) + 2H_2O(l) \rightarrow Ca(OH)_2(aq) + H_2(g)$$

- Strontium and barium react similarly, the reaction of barium being vigorous and giving a colourless solution of barium hydroxide — the most soluble of the group 2 hydroxides.

Reaction of group 2 oxides with water

The oxides of group 2 metals react with water to give hydroxides. For example:

$$MgO(s) + H_2O(l) \rightarrow Mg(OH)_2(aq) \text{ and (s)}$$

Magnesium hydroxide is sparingly soluble. The solubility of the hydroxides increases down group 2.

The reaction is that of the basic oxide ion with water:

$$O^{2-} + H_2O \rightarrow 2OH^-$$

Reactions of group 2 oxides and hydroxides with dilute acids

All the group 2 oxides are basic. The reaction of magnesium oxide is typical:

$$MgO(s) + 2H^+(aq) \rightarrow Mg^{2+}(aq) + H_2O(l)$$

Reaction of the other group 2 oxides with sulfuric acid produces a coating of the insoluble metal sulfate, which prevents further reaction.

The basic hydroxides react similarly with acid:

$$Mg(OH)_2(s) + 2H^+(aq) \rightarrow Mg^{2+}(aq) + 2H_2O(l)$$

Solubility of group 2 hydroxides and sulfates

You need only know the trends in solubility of the group 2 hydroxides and sulfates — you are not required to explain them. The solubility values are given in moles of solute per 100 g water at 25°C.

Hydroxides

The solubility of the hydroxides increases with increasing atomic number of the cation.

Compound	$Mg(OH)_2$	$Ca(OH)_2$	$Sr(OH)_2$	$Ba(OH)_2$
Solubility	2.00×10^{-5}	1.53×10^{-3}	3.37×10^{-3}	1.50×10^{-2}

Sulfates

The solubility of the sulfates falls with increasing atomic number of the cation. In practice, only magnesium sulfate is noticeably soluble.

Compound	$MgSO_4$	$CaSO_4$	$SrSO_4$	$BaSO_4$
Solubility	1.83×10^{-1}	4.66×10^{-3}	7.11×10^{-5}	9.43×10^{-7}

Thermal stability

Nitrates of groups 1 and 2

Nitrates decompose on heating to give either the metal nitrite and oxygen (sodium to caesium in group 1):

$$2NaNO_3 \rightarrow 2NaNO_2 + O_2$$

or nitrogen dioxide, oxygen and the metal oxide (lithium in group 1, magnesium to barium in group 2):

$$2LiNO_3 \rightarrow Li_2O + 2NO_2 + \tfrac{1}{2}O_2$$

$$M(NO_3)_2 \rightarrow MO + 2NO_2 + \tfrac{1}{2}O_2$$

where M is a group 2 metal.

The ease of decomposition is related to the polarising power of the cation. Thus, magnesium nitrate contains the small, polarising cation Mg^{2+}, which polarises the nitrate ion and makes decomposition easier. The larger Ba^{2+} cation in barium nitrate causes less polarisation and the decomposition is more difficult. Group 1 nitrates decompose with difficulty to the nitrite, except lithium nitrate, which has the smallest and therefore the most polarising cation in this group (Li^+).

Carbonates of group 2

Group 2 carbonates decompose on heating to give the metal oxide and carbon dioxide. Decomposition becomes more difficult down the group, as the cation gets bigger and less polarising. The magnesium ion (Mg^{2+}) is the smallest and therefore the most polarising cation in group 2, so magnesium carbonate is thermally the least stable.

The decomposition of calcium carbonate is important in the blast furnace for the production of iron and other metals and in the manufacture of Portland cement.

$$CaCO_3(s) \rightarrow CaO(s) + CO_2(g)$$

In a closed system this reaction is in equilibrium.

Flame tests

If the atoms or ions of some group 1 or 2 elements are heated in a Bunsen flame, they emit light of distinctive colours. The flame excites electrons to higher energy orbitals. When the electrons fall back down, they emit light of characteristic wavelength.

Metal	Lithium	Sodium	Potassium	Calcium	Strontium	Barium
Colour	Crimson red	Yellow	Lilac	Orange-red or brick red	Carmine red	Apple green

The reds of lithium and strontium are difficult to tell apart. In a qualitative analysis other tests would be needed. The flame colours can be used quantitatively to determine ion concentrations (e.g. in blood or urine) using a flame photometer.

Acid–base titrations: volumetric analysis

Strong bases are often hydroxides of alkali metals (group 1) or alkaline earth metals (group 2). The hydrohalic acids are strong acids. Acid–base titrations are among the most common volumetric processes. As with all other practical skills in chemistry, proper technique is essential. In quantitative analysis poor technique will produce inaccurate results.

Techniques used in volumetric analysis

Volumetric analysis, or titration, involves the reaction of two solutions. The volume and the concentration are known for one solution. For the other solution, only the volume is known. The apparatus used is the burette, the pipette and the graduated flask. Accurate results require careful technique in the use of each of these.

Standard solutions

The solution for which the concentration is accurately known is the **standard solution**. The concentration may have been found by a previous titration or by weighing the solute and making it to a known volume. A solution made by weighing is a **primary standard solution**.

Not all substances are suitable for use as primary standards. If a substance is to be weighed accurately enough for use in a standard solution, the following criteria must be met:
- The substance must be commercially available in a high state of purity or must be easily purified (e.g. by recrystallisation).
- The substance must not be volatile, otherwise some will be lost during weighing.
- The substance must not react with oxygen, carbon dioxide or water, otherwise it cannot be weighed in air or dissolved in water.

Examples of primary standards are:
- strong acid — sulfamic acid (NH_2SO_3H), a white solid
- strong base — anhydrous sodium carbonate (Na_2CO_3), a white solid
- oxidising agent — potassium dichromate(VI) ($K_2Cr_2O_7$), a bright orange solid
- reducing agent — ammonium iron(II) sulfate heptahydrate [$(NH_4)_2SO_4.FeSO_4.7H_2O$], a pale green solid

Preparing a standard solution

Weighing the solute accurately requires cleanliness and care. It is assumed that you are using an electronic balance with a tare facility, where the reading can be made zero by pressing a button.

- Make sure the balance pan is clean and dry. Place the weighing bottle on the pan and tare the balance.
- Add a suitable amount of the solid to the bottle. Do this by taking the bottle off the pan and adding the solid away from the balance, so any spillage does not fall on the pan. Try to avoid spillage.
- When you have the required amount, write its value down immediately in your notebook.
- Transfer the solid to the graduated flask using a funnel. Wash out the weighing bottle into the funnel using a wash bottle and rinse all the solid into the flask. Add about 50 cm^3 of pure water and shake to dissolve.
- Some materials may have large crystals or may set solid when wetted. In these cases the solid is transferred to a beaker, the weighing bottle washed into the beaker and about 50 cm^3 of water added. The solution is stirred with a glass rod until the solid has dissolved. The solution is transferred completely to the gradu-ated flask by pouring it down the same glass rod into a funnel and by washing any remaining solution off the beaker and the glass rod into the funnel.
- Whichever method is used, add pure water so that the lower level of the meniscus is on the mark. Stopper the flask and mix thoroughly by inverting and shaking the flask vigorously five or six times. Simple shaking is not enough; most serious errors in volumetric analysis can be traced to poor mixing of the solution in the gradu-ated flask.

Using a pipette

A glass bulb pipette delivers the volume stated on it within acceptable limits only if it is used as intended. The use of a pipette filler is obligatory. Apart from the hazards associated with getting some chemicals into your mouth, it avoids contamination of solutions with saliva. In safety questions in examinations, credit is not given for answers referring to the use of lab coats, safety glasses or pipette fillers, since all are assumed to be part of routine good practice.

- Using a pipette filler, draw a little of the solution to be pipetted into the pipette and use this to rinse the pipette; discard these rinsings.
- Fill the pipette to about 2–3 cm above the mark. Pipette fillers are difficult to adjust accurately, so quickly remove the filler and close the pipette with your forefinger (not thumb). Release the solution until the bottom of the meniscus is on the mark.
- Immediately transfer the pipette to the conical flask in which you will do the titra-tion and allow the solution to dispense under gravity. Under no circumstances blow it out. Good analysis requires patience. When all the solution appears to have been dispensed, wait 15–20 seconds, touch the tip of the pipette under the surface of the liquid and then withdraw the pipette. It is calibrated to allow for the liquid remaining in the tip.

Using a burette

A burette dispenses solutions accurately only if used correctly — there are two partic-ular pitfalls that can cause serious inaccuracies.

- Making sure that the tap is shut, add about 10 cm^3 of the solution you intend to use into the burette and rinse it out with this solution, not forgetting to open the tap and rinse the jet.

- Close the tap and fill the burette. A small funnel is helpful, but be careful not to overfill it, otherwise you will be rinsing the outside of the burette as well. Remove the funnel; titrating with a funnel in the burette can lead to serious error if a drop of liquid in the funnel stem falls into the burette during the titration.
- Bring the meniscus on to the scale by opening the tap over a suitable receptacle. There is no particular reason to bring the meniscus exactly to the zero mark.
- Make sure that the burette is full to the tip of the jet. Failure to ensure this is another source of serious error.
- Titration is a two-handed process. Add a suitable indicator to the solution in the conical flask, then swirl this under the burette with your right hand while manipulating the burette tap with your left. Your thumb and forefinger should encircle the burette. This feels awkward at first, but after practice becomes natural and gives good control of the tap.
- Add the titrant (the solution in the burette) slowly, swirling the flask all the time. As the end point is approached, the indicator will change colour more slowly. The titrant should be added drop by drop near to the end point — your aim is to make it change with the addition of one drop of titrant. Wait a few moments before reading the burette — this is to allow the solution time to drain down the walls of the burette.
- It is poor practice to run the titrant rapidly into the conical flask once you have an idea of the required volume. In some cases, particularly with redox titrations (not acid–base) involving potassium manganate(VII), rapid addition can cause precipitation reactions that do not necessarily reverse when the solution is swirled.
- The titration is repeated until you have three concordant titres, that is, volumes that are similar. This means within 0.2 cm^3 or better if you have been careful. Taking the mean of three titres that differ by 1 cm^3 or more cannot give accurate answers.

Indicators

The two common indicators for acid–base titrations are methyl orange and phenolphthalein. Methyl orange is red below pH 3 and yellow above about pH 6. Phenolphthalein is colourless below about pH 8 and pink (or magenta) above pH 10. Many acid–base titrations produce a large change in pH from about 1.5 or 2 to 12 or 13 (if base is being added to acid) within a few drops of the end point volume, so either indicator can be used. However, a strong acid–weak base titration needs methyl orange and a strong base–weak acid titration needs phenolphthalein. (The theory of indicators is covered at A2.) Another example of an indicator is bromothymol blue, which is yellow below pH 6 and blue above pH 8.

Volumetric calculations

It is essential to lay out calculations clearly. They must be easy to read, make sense and be realistic in terms of significant figures used. Adopting the following principles will improve your understanding of volumetric calculations.

- The whole calculation uses numbers of moles rather than some formula which might be mis-remembered or misapplied.
- Use units throughout.
- The word 'amount' is used in its technical chemical sense, i.e. a quantity of moles.
- The molar mass of a compound is calculated by writing out each atomic mass explicitly. Incorrect molar masses cost candidates many exam marks because the examiner cannot tell whether the error is arithmetical or chemical. (In your everyday work you could look the value up in a data book. If you do this, *tell* the examiner that you have done so.)
- The dimensionless quantity *relative* molecular mass is not used. No calculation uses this quantity.

Example

25.0 cm^3 of a solution of sodium carbonate was titrated with hydrochloric acid solution of concentration 0.108 mol dm^{-3} using methyl orange indicator. The volume needed was 27.2 cm^3. Find the concentration of the sodium carbonate solution in mol dm^{-3} and in g dm^{-3} of the anhydrous salt.

The reaction is:

$$Na_2CO_3 + 2HCl \rightarrow 2NaCl + H_2O + CO_2$$

Amount of hydrochloric acid used = 0.0272 dm^3 × 0.108 mol dm^{-3} = 0.00294 mol

Thus amount of sodium carbonate = 1/2 × 0.00294 mol = 0.00147 mol in 25.0 cm^3

Thus concentration of sodium carbonate solution = 0.00147 mol ÷ 0.025 dm^3
= 0.0588 mol dm^{-3}

The molar mass of anhydrous sodium carbonate is:
[(2 × 23) +12 + (3 × 16)] g mol^{-1} = 106 g mol^{-1}

The concentration of the sodium carbonate is therefore:
0.0588 mol dm^{-3} × 106 g mol^{-1} = 6.233 g dm^{-3}

Group 7 (halogens)

Physical properties of the elements

Element	State at room temperature	Melting temperature/K	Boiling temperature/K
Fluorine	Pale yellow gas	53.5	85.0
Chlorine	Greenish-yellow gas	172	239
Bromine	Brown volatile liquid	266	332
Iodine	Dark grey lustrous solid; the vapour is purple	387	457

Iodine does not sublime on heating under normal pressure. It is volatile even as a solid and has a low heat capacity. Heating with a Bunsen burner causes such a rapid temperature rise that the liquid phase is not usually seen. With gentle heating the liquid phase is obvious.

Chlorine and bromine dissolved in water or in organic solvents give pale green and orange or yellow solutions respectively. Iodine is purple when dissolved in liquids that do not contain oxygen in the molecule (e.g. hexane or benzene). In aqueous or alcoholic solution, iodine is brown. It is not very soluble in water. Addition of potassium iodide to aqueous iodine gives a brown solution of potassium triiodide (KI_3), often (but inaccurately) labelled 'iodine solution':

$$I_2(aq) + KI(aq) \rightleftharpoons KI_3(aq)$$

Oxidising reactions of the halogens

Halogens are oxidising agents. Their oxidising power decreases in the order chlorine > bromine > iodine. Oxidising agents are reduced (gain electrons) when they react, so the smaller the halogen atom, the more strongly attracted the electron gained will be.

Reactions with metals

On heating, metals react with halogens to form halides. The following reactions are typical of the halogens. Unless otherwise shown, the reactions with bromine and iodine are the same as those with chlorine.

$$2Na(l) + Cl_2(g) \rightarrow 2NaCl(s)$$

$$Ca(s) + Cl_2(g) \rightarrow CaCl_2(s)$$

Iron can form Fe^{2+} or Fe^{3+} ions. The reaction of iron with chlorine gives iron(III) chloride:

$$2Fe(s) + 3Cl_2(g) \rightarrow 2FeCl_3(s)$$

Reaction of iron with the less powerful oxidising agent iodine gives iron(II) iodide:

$$Fe(s) + I_2(g) \rightarrow FeI_2(s)$$

Reactions with non-metals

Non-metals react with halogens to give covalent compounds. Usually the chloro compounds are the only ones of significance, except for the compounds with hydrogen. Hydrogen burns in an atmosphere of chlorine or bromine. Mixtures of the gases explode if heated.

$$H_2(g) + Cl_2(g) \rightarrow 2HCl(g)$$

$$H_2(g) + Br_2(g) \rightarrow 2HBr(g)$$

The reaction with iodine is in equilibrium at 400°C. This was one of the first equilibrium reactions to be studied in detail.

$$H_2(g) + I_2(g) \rightleftharpoons 2HI(g)$$

White phosphorus, on heating with chlorine, initially gives phosphorus trichloride. This product reacts with excess chlorine to form phosphorus pentachloride:

$$P_4(l) + 6Cl_2(g) \rightarrow 4PCl_3(l)$$

$$PCl_3(l) + Cl_2(g) \rightleftharpoons PCl_5(g)$$

Reactions of aqueous halogens

In aqueous solution, chlorine oxidises:

- iron(II) ions to iron(III) ions:
  $$2Fe^{2+}(aq) + Cl_2(aq) \rightarrow 2Fe^{3+}(aq) + 2Cl^-(aq)$$
- sulfite ions to sulfate ions:
  $$SO_3^{2-}(aq) + Cl_2(aq) + H_2O(l) \rightarrow SO_4^{2-}(aq) + 2H^+(aq) + 2Cl^-(aq)$$
- nitrite ions to nitrate ions:
  $$NO_2^-(aq) + Cl_2(aq) + H_2O(l) \rightarrow NO_3^-(aq) + 2H^+(aq) + 2Cl^-(aq)$$

Bromine reacts similarly but more slowly. Iodine does not necessarily react in the same way — for example, it does not oxidise iron(II) ions.

Oxidation of halide ions by halogens: displacement reactions

Chlorine oxidises (displaces) both bromide and iodide ions. Bromine, a less powerful oxidising agent than chlorine, oxidises iodide ions. The first reaction is used to manufacture bromine from seawater:

$$Cl_2(aq) + 2Br^-(aq) \rightarrow 2Cl^-(aq) + Br_2(aq)$$

$$Cl_2(aq) + 2I^-(aq) \rightarrow 2Cl^-(aq) + I_2(aq)$$

$$Br_2(aq) + 2I^-(aq) \rightarrow 2Br^-(aq) + I_2(aq)$$

Disproportionation reactions

Disproportionation is the simultaneous oxidation and reduction of an atom in a reaction. The halogens form oxyanions in which the halogen has a positive oxidation state. Thus chlorate(I) ions (OCl$^-$) contain chlorine in the +1 state and chlorate(V) ions (ClO$_3^-$) have chlorine in the +5 state.

Reaction of chlorine with water or sodium hydroxide solution causes it to disproportionate.

- chlorine with water:
  $$Cl_2(aq) + H_2O(l) \rightarrow HOCl(aq) + HCl(aq)$$
- chlorine with cold dilute NaOH solution:
  $$Cl_2(aq) + 2NaOH(aq) \rightarrow NaCl(aq) + NaOCl(aq) + H_2O(l)$$
- chlorine with hot concentrated NaOH solution:
  $$3Cl_2(aq) + 6NaOH(aq) \rightarrow 5NaCl(aq) + NaClO_3(aq) + 3H_2O(l)$$

Chlorate(I) ions disproportionate on heating in solution to give chlorate(V) and chloride:

$$3OCl^-(aq) \rightarrow ClO_3^-(aq) + 2Cl^-(aq)$$

The halogens behave identically in these reactions.

Iodine titrations

Solutions of copper(II) salts give a brown mixture containing iodine and copper(I) iodide when added to solutions of iodides. Addition of sodium thiosulfate ($Na_2S_2O_3$) solution decolorises the iodine and leaves pinkish-cream copper(I) iodide as a precipitate.

$$2Cu^{2+}(aq) + 4I^-(aq) \rightarrow 2CuI(s) + I_2(aq)$$

$$2S_2O_3^{2-}(aq) + I_2(aq) \rightarrow 2I^-(aq) + S_4O_6^{2-}(aq)$$

Use of iodine in volumetric analysis

There are several volumetric techniques that make use of a compound's ability to oxidise iodide ions to iodine. The iodine produced is estimated by titration with a standard solution of sodium thiosulfate ($Na_2S_2O_3$). Sodium thiosulfate is a primary standard.

- A known volume of the oxidising agent is reacted with an excess of potassium iodide, liberating iodine.
- The liberated iodine is titrated with a standard solution of sodium thiosulfate. The reaction is:

$$2S_2O_3^{2-} + I_2 \rightarrow S_4O_6^{2-} + 2I^-$$

- The end point (when all the iodine has reacted) is reached when the solution changes from pale yellow to colourless. To make this easier to see, starch can be added when the solution is pale yellow, *but not earlier*. The colour change is then blue-black to colourless.
- If starch is added too soon, a precipitate of 'starch iodide' forms, which does not dissolve as the sodium thiosulfate is added. This would make the titre obtained too low.

The above technique can be used in the following titrations.

Determining the amount of copper in brass

- A weighed sample of brass is dissolved in the minimum amount of concentrated nitric acid and the solution is made to a known volume (e.g. 250 cm^3).

$$Cu + 4HNO_3 \rightarrow Cu(NO_3)_2 + 2NO_2 + 2H_2O$$

- A 25.0 cm^3 sample of this solution is pipetted into a conical flask and neutralised (to remove excess nitric acid) with sodium carbonate solution. An excess of potassium iodide solution is added. The copper(II) ions are reduced to copper(I) and iodine is liberated:

$$2Cu^{2+}(aq) + 4I^-(aq) \rightarrow 2CuI(s) + I_2(aq)$$

- The iodine is titrated with a standard solution of sodium thiosulfate as detailed above. Since $2S_2O_3^{2-} \equiv I_2$, and $I_2 \equiv 2Cu^{2+}$ from the equations for the reactions, it is clear that $S_2O_3^{2-} \equiv Cu^{2+} \equiv Cu$, so the amount of copper in the sample can be found.

Determining the percentage purity of a sample of potassium iodate(v)

- A weighed sample of potassium iodate(v) (KIO_3) is dissolved in the minimum amount of concentrated nitric acid and the solution is made to a known volume (e.g. 250 cm^3).

- A 25.0 cm³ sample of this solution is pipetted into a conical flask and acidified with sulfuric acid. An excess of potassium iodide solution is added. The iodate(v) ions oxidise iodide ions to iodine in acidic solution:

$$IO_3^-(aq) + 5I^-(aq) + 6H^+(aq) \rightarrow 3I_2(aq) + 3H_2O(l)$$

- The iodine is titrated with a standard solution of sodium thiosulfate as detailed above. Since $2S_2O_3^{2-} \equiv I_2$, and $3I_2 \equiv IO_3^-$ from the equations for the reactions, it is clear that $6S_2O_3^{2-} \equiv IO_3^-$, so the amount of iodate(v) in the sample can be found.

Reaction of halide salts

Reaction with concentrated sulfuric acid

Sulfuric acid is a stronger acid than the hydrogen halides (HX, X = Cl, Br or I). The hydrogen halides are also gases, so if sulfuric acid is added to a potassium halide, HX is liberated, since sulfuric acid donates a hydrogen ion to the halide ion. For example, with KCl:

$$KCl(s) + H_2SO_4(l) \rightarrow KHSO_4(s) + HCl(g)$$

Steamy fumes of HCl are given off.

In the case of bromide and iodide salts, the HBr or HI liberated is a strong enough reducing agent to be oxidised by sulfuric acid, so little HX is obtained. Bromides are oxidised to bromine and the sulfuric acid is reduced to sulfur dioxide. Orange-brown fumes are evolved:

$$KBr(s) + H_2SO_4(l) \rightarrow KHSO_4(s) + HBr(g)$$

$$2HBr(g) + H_2SO_4(l) \rightarrow Br_2(g) + SO_2(g) + 2H_2O(l)$$

Iodide is an even stronger reducing agent and can reduce sulfuric acid in three ways. The reaction produces purple fumes of iodine, a smell of rotten eggs from the hydrogen sulfide and a brown sludge in the test-tube, since some of the iodine forms brown KI_3 with excess KI. State symbols do not help with this series of reactions.

$$KI(s) + H_2SO_4(l) \rightarrow KHSO_4(s) + HI(g)$$

$$2HI + H_2SO_4 \rightarrow I_2 + SO_2 + 2H_2O$$

$$6HI + H_2SO_4 \rightarrow 3I_2 + S + 4H_2O$$

$$8HI + H_2SO_4 \rightarrow 4I_2 + H_2S + 4H_2O$$

Reaction with silver nitrate

The test solution is made acidic with nitric acid, which decomposes carbonates or sulfites that would interfere with the test. Silver nitrate solution is then added:

$Ag^+(aq) + Cl^-(aq) \rightarrow AgCl(s)$ White precipitate

$Ag^+(aq) + Br^-(aq) \rightarrow AgBr(s)$ Cream precipitate

$Ag^+(aq) + I^-(aq) \rightarrow AgI(s)$ Yellow precipitate

The precipitates are then treated with ammonia solution. Silver chloride dissolves in dilute ammonia to give a colourless solution:

$$AgCl(s) + 2NH_3(aq) \rightarrow [Ag(NH_3)_2]^+(aq) + Cl^-(aq)$$

Silver bromide dissolves in concentrated ammonia to give a colourless solution:

$$AgBr(s) + 2NH_3(aq) \rightarrow [Ag(NH_3)_2]^+(aq) + Br^-(aq)$$

Silver iodide is too insoluble to react with ammonia.

Silver halides are decomposed by light. Photochromic lenses contain silver chloride, which dissociates to form silver in the glass. When the light source is removed, the silver recombines with the chlorine. Photographic emulsions contain silver chloride or bromide. Exposure to light makes it easier to reduce the silver halide to silver in the developing bath than the unexposed compound.

Hydrogen halides

Dry hydrogen halide gases are not acidic. In aqueous solution the hydrogen halides form strongly acidic solutions. The formation of the hydrated H_3O^+ and X^- ions (X = Cl, Br or I) liberates enough energy to compensate for the breaking of the H–X bond.

$$HX(g) + H_2O(l) \rightarrow H_3O^+(aq) + X^-(aq)$$

The solutions are typical strong acids and react with bases to form salts. Thus with ammonia:

$$HX(aq) + NH_3(aq) \rightarrow NH_4X(aq)$$

Fluorine and astatine

The chemistry of the halogens shows a moderately clear trend from fluorine to iodine, so knowledge of the chemistry of chlorine, bromine and iodine enables the chemistry of the other halogens, fluorine and astatine, to be predicted.

Fluorine

It would be expected that, based on the chemistry of the other halogens, fluorine has the following properties:

- It is a powerful oxidising agent — the most powerful, in fact. Fluoride ions cannot be oxidised by any other oxidising agent, so fluorine can only be made by electrolysis.
- Fluorine is the most electronegative element.
- It is also the most reactive element and forms compounds with all other elements, apart from the first three noble gases (He, Ne and Ar).
- Fluorine reacts with metals to give ionic compounds; it reacts with non-metals to give covalent compounds.
- Fluorine usually brings out the highest possible oxidation state in the element with which it combines — for example, in reaction with sulfur it forms SF_6, whereas reaction of sulfur with chlorine gives SCl_4.

The high reactivity of fluorine is a result of:

- the weakness of the F–F bond; it is so short that non-bonding electrons in the two atoms repel one another and weaken the bond, which is not much stronger than

that in iodine (bond enthalpies: F–F = 158 kJ mol^{-1}, I–I = 151 kJ mol^{-1})
- the small size of the fluoride ion, resulting in ionic compounds with high lattice energies
- the highly exothermic formation of hydrated fluoride ions (though high lattice energies mean that many fluorides are not water-soluble)
- strong covalent bonds formed with non-metallic elements because the bond is short (bond enthalpies: H–F = 568 kJ mol^{-1}, H–Cl = 462 kJ mol^{-1})

The high reactivity of fluorine means that some of its reactions differ from those of the other halogens. For example, fluorine reacts with alkanes to give hydrogen fluoride and carbon, rather than fluoroalkanes. The organic chemistry of fluorine is a specialist study.

Astatine
Astatine is radioactive and occurs in vanishingly small traces in the ore uranite. Its longest-lived isotope is astatine-211 (^{211}At), which has a half-life ($t_{1/2}$) of 8.3 hours. Microgram quantities have been synthesised, but the element has never been seen. Tracer experiments using carrier compounds of iodine suggest that astatine is like iodine, but more metallic and would appear as a dark-coloured solid with a metallic sheen. Astatide ions would be easily oxidised to the element and its compounds would show a high degree of covalent character.

Kinetics

The factors that control the rate of a chemical reaction include:
- the concentration of the reagents (for reactions in solution)
- the temperature of the reaction system
- the pressure (for reactions in the gas phase)
- the surface area of any solid reagents
- the presence of a catalyst

Collision theory

Chemicals cannot react unless they collide. The theory of reaction rates is therefore called **collision theory**. When thinking about reaction rates, you should do so on a molecular level and try to imagine the collisions occurring.

The important factors are:
- the number of collisions per unit time — the **collision frequency**
- the energy with which the particles collide — the **collision energy**
- the **orientation** in which the particles collide — particularly important for large molecules

Only a small proportion of the collisions that occur in a reaction system are successful, i.e. lead to the formation of products.

Collision frequency

Collision frequency increases with concentration in a liquid system, or with an increase in pressure in a gaseous one. In each case, the distance between colliding species is reduced, so there is less distance to travel before encountering another molecule.

Collision frequency increases with surface area for a solid reagent. In this case, the increased area raises the probability of a molecule in the other phase (gas or liquid) colliding with the solid.

Collision frequency increases with temperature. The molecules are moving faster and so travel the necessary distance more quickly.

Collision energy

The minimum collision energy needed for particles to react is called the **activation energy (E_a)**. Particles that collide with an energy greater than E_a react if their orientation is correct.

Increasing the temperature increases the proportion of particles that collide with energies greater than E_a. The effect of an increase in temperature on collision energy is more important than the effect on collision frequency. The collision frequency rises roughly linearly with rising temperature, but the increase in the number of particles with collision energies above E_a is exponential.

Orientation

Particles must collide in such a way that their reactive parts come into contact.

Maxwell–Boltzmann distribution of molecular energies

The graphs below show the distribution of energy between molecules in a gas at temperature T_1 and at a higher temperature T_2.

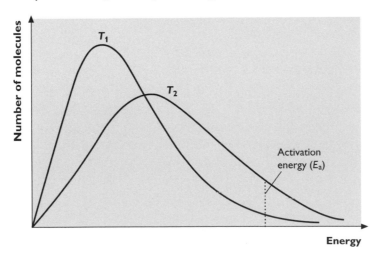

There are several points to note about these graphs.
- Both graphs start at the origin.
- The distribution is skewed.
- The graph does *not* meet the x-axis at high energies.
- The higher temperature graph has its peak at higher energy, but the peak is lower than the graph for T_1. This is because the area under the two graphs must be the same, since this area is proportional to the total number of molecules.

The mathematical expression leading to the graph was originally derived for a single gas. Its use for reaction mixtures, whether in the gaseous or liquid phases, is a sensible extension of the original idea but is diagrammatic rather than quantitatively precise.

Activation energy is shown on the diagram. This is explained below.

Activation energy

The minimum collision energy needed for particles to react is called the activation energy (E_a). Particles that collide with energy greater than E_a react if their orientation is correct.

Activation energy is represented above on the Maxwell–Boltzmann distribution. The area to the right of E_a represents the number of molecules that possess the activation energy or more and that therefore could react. However, remember that it is the overall collision energy that matters.

At the higher temperature, the area to the right of E_a increases, as does the number of successful collisions. Therefore, the rate increases. The E_a is well to the right of the peak. Therefore, the proportion of successful collisions is small.

Examination questions often ask for reference to a diagram such as that above — make sure you refer to the diagram in your answer.

Catalysts

Catalysts change the mechanism of a reaction to one having a lower value of E_a. This means that the proportion of successful collisions at a given temperature increases and therefore the rate increases. Do not say that 'a catalyst lowers the activation energy' because this will get you no marks in the examination. You need to state that the *mechanism* changes.

At least one of the reactants must combine with the catalyst as an initial step. The complex formed then reacts with another reactant to give the products and regenerate the catalyst. The reaction profile for a catalysed reaction therefore has at least two humps, with an intermediate reactant–catalyst complex. This is shown in the diagram below.

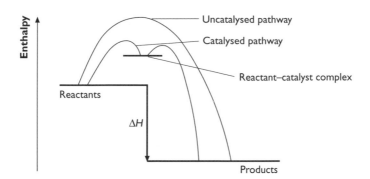

Thermodynamic and kinetic stability

Methane and oxygen react only if ignited, to give carbon dioxide, water and heat:

$$CH_4 + 2O_2 \rightarrow CO_2 + 2H_2O \qquad \Delta H = -890 \text{ kJ mol}^{-1}$$

The fact that carbon dioxide and water are at a lower energy than methane and oxygen means that carbon dioxide and water are **thermodynamically stable** with respect to methane and oxygen. The idea of thermodynamic stability is comparative — you must say that one mixture is stable *with respect to another*.

The reaction between methane and oxygen is immeasurably slow at room temperature. This is because the activation energy is sufficiently high for there to be virtually no molecules that collide with this energy at room temperature. Methane and oxygen are therefore **kinetically stable** with respect to carbon dioxide and water.

Thermodynamic stability is a question of energetics, i.e. ΔH; kinetic stability is a question of kinetics, i.e. E_a.

Chemical equilibria

Dynamic equilibrium

A reaction in dynamic equilibrium:
- is undergoing no net change, so the concentration of each substance is constant
- has a composition that can be approached starting from the reactants or the products
- is one in which the forward reaction (conventionally written from left to right) is happening at the same rate as the reverse reaction

A homogeneous equilibrium is one in which all the reagents and products are in the same phase. A phase has a clear boundary and it is not the same as the state of a material. A mixture of oil and water is entirely in the liquid state, but because there are two layers it is a two-phase system.

An example of a liquid-phase equilibrium system is that involving ethanol, ethanoic acid, ethyl ethanoate and water:

$$CH_3CH_2OH + CH_3COOH \rightleftharpoons CH_3COOCH_2CH_3 + H_2O$$

This is an esterification reaction.

An example of a gas-phase equilibrium is the Haber process for the manufacture of ammonia:

$$N_2(g) + 3H_2(g) \rightleftharpoons 2NH_3(g)$$

Change of conditions

Effect of change in concentration

The esterification reaction is used as an example. If the concentration of either ethanol or ethanoic acid (or, in general, the substances on the left-hand side of the equation) is increased, the forward reaction occurs more rapidly until (new) equilibrium concentrations are established. The new equilibrium position will have a greater concentration of ethyl ethanoate and water (or, in general, the products — the substances on the right-hand side of the equation) than before. Thus, the yield is higher.

Reducing the concentration of the substances on the right-hand side, by removing them from the reaction vessel, has the same effect of increasing the yield. A similar argument can be put forward for the reaction 'moving to the left', i.e. for the concentrations of substances on the left to increase if the concentration of the substances on the right-hand side is increased.

There is an infinite number of compositions that can give rise to equilibrium for a particular system — the equilibrium composition is not unique. Apart from anything else there is an infinite variety of compositions of the reactants that you could use to produce an equilibrium system.

Effect of change in pressure

The Haber process equilibrium is used as an example. Increasing the total equilibrium pressure moves the equilibrium composition towards the side with the smaller number of moles, i.e. the side with the smaller volume. In this case, the amount of ammonia would increase. A similar argument can be put forward for a decrease in pressure, which would move the equilibrium composition towards the reactants.

Effect of change in temperature

The effect of a change in the equilibrium temperature depends on whether the reaction, *defined in the forward direction* (left to right, as written), is exothermic or endothermic. (You need to state clearly which direction you are talking about in answers to examination questions.) The synthesis of ammonia is exothermic in the forward direction:

$$N_2(g) + 3H_2(g) \rightarrow 2NH_3(g) \qquad \Delta H = -92 \text{ kJ mol}^{-1}$$

An increase in the equilibrium temperature moves the equilibrium position in the endothermic direction, i.e. to the left, so the yield of ammonia falls at higher temperatures.

A similar argument can be made for endothermic reactions, where an increase in the (equilibrium) temperature moves the equilibrium position to the right. For example, the equilibrium between dinitrogen tetroxide (N_2O_4, colourless) and nitrogen dioxide (NO_2, brown) is endothermic from left to right, so increasing the temperature of the system makes the gas mixture a darker brown.

Remember that the temperatures are equilibrium temperatures, which are externally imposed, and therefore it is assumed that any change in equilibrium composition is not going to be able to affect that temperature. For this to be true, the surroundings are assumed to have such a large heat capacity that absorption or evolution of heat energy by the system does not affect the temperature of the surroundings. The equilibrium mixture is said to be in a thermostat.

Avoid Le Chatelier

You are likely to have come across Le Chatelier's principle, which states that changing the conditions under which an equilibrium is set up 'changes the composition in such a way as to tend to oppose the change'. Much confusion would be avoided if Le Chatelier were consigned to history. You will discover at A2 that it is an unnecessary idea in any case. It causes confusion because of the following:

- It leads some students to write about equilibria in a way that suggests that the reaction mixture can think — 'the equilibrium tries to reduce the temperature...'. This approach is poor and it matters because it obscures the true behaviour of equilibrium systems, which reach equilibrium because of the thermodynamics of the system. The composition of the system at equilibrium is that which minimises the energy of the system.
- It leads some students to believe that when the equilibrium composition changes, for example by raising the temperature, the system brings the temperature back down again to what it was. If it did do this, the composition would still be what it was initially. The temperature, the composition and the value of ΔH for the reaction are inextricably intertwined. The two different temperatures are *externally imposed* and represent two different *equilibrium* states with different compositions.

Note that arguments based on Le Chatelier's principle no longer earn marks in examinations.

Organic chemistry

Alcohols

Alcohols are organic compounds containing one (or more) hydroxyl groups (–OH) as the functional group. Note that –OH is not 'hydroxide'.

There are three types of alcohol $C_nH_{2n+1}OH$ that have one –OH group. Each type is shown in the table below, together with the names and different representations for the isomers of the first four members of the alcohol homologous series. Skeletal formulae are not generally used where a compound has fewer than three carbon atoms. In each case, R, R' and R" are organic groups that can be different or the same. For primary alcohols alone, R can be H (giving methanol).

Primary	Secondary	Tertiary
Methanol (CH_3OH) 	—	—
Ethanol (CH_3CH_2OH) 	—	—
Propan-1-ol ($CH_3CH_2CH_2OH$) 	Propan-2-ol ($CH_3CH(OH)CH_3$) 	—
Butan-1-ol ($CH_3CH_2CH_2CH_2OH$) 	Butan-2-ol ($CH_3CH_2CH(OH)CH_3$) 	2-methylpropan-2-ol ($(CH_3)_3COH$)

Primary	Secondary	Tertiary
2-methylpropan-1-ol ($CH_3CH(CH_3)CH_2OH$)	—	—

Reactions of alcohols

Combustion

All alcohols burn in air to give carbon dioxide and water. Because they contain an oxygen atom, they burn with a cleaner flame than hydrocarbons do. Alcohols become less easy to burn as the carbon chain gets longer because they become less volatile.

$$CH_3CH_2OH + 3O_2 \rightarrow 2CO_2 + 3H_2O$$

The general equation for the combustion of alcohols is:

$$C_nH_{2n+1}OH + 3n/2O_2 \rightarrow nCO_2 + (n + 1)H_2O$$

Ethanol can be produced by the fermentation of sugar cane and added to petrol for motor fuel. The ethanol is therefore a renewable fuel, but any land given over to growing fuel cannot be used for growing food.

Reaction with sodium

Alcohols react with sodium as they are weak acids. The hydrogen of the alcohol is reduced to hydrogen gas. The reaction is slow for alcohols larger than C_3. The alkoxide ions RO^- produced are strong bases and are good nucleophiles, making them useful reagents in synthetic organic chemistry. With ethanol the products are sodium ethoxide and hydrogen:

$$2CH_3CH_2OH + 2Na \rightarrow 2CH_3CH_2O^-Na^+ + H_2$$

Reaction with halogenating agents

Alcohols can be converted into halogenoalkanes in several ways, which work for all three types of alcohol.

Using phosphorus pentachloride

Solid phosphorus pentachloride (PCl_5) reacts readily with alcohols at room temperature to give the chloroalkane and HCl gas, which is emitted as steamy fumes.

$$CH_3CH_2OH + PCl_5 \rightarrow CH_3CH_2Cl + HCl + POCl_3$$

Test for the –OH group

Reaction with PCl_5 can be used as a test for presence of an –OH (hydroxyl) group. It is not specific to alcohols and the –OH groups in carboxylic acids give the same reaction, but these acids would also affect indicators, which alcohols do not.

Using a mixture of sodium bromide and 50% sulfuric acid

The alcohol is heated under reflux with sodium bromide and 50% sulfuric acid. The method can be adapted to make chlorides using NaCl, but not for iodides, since sulfuric acid oxidises iodides to iodine.

$$CH_3CH_2OH + NaBr + H_2SO_4 \rightarrow CH_3CH_2Br + NaHSO_4 + H_2O$$

Using a mixture of iodine and moist red phosphorus

Warming alcohols with iodine and moist red phosphorus produces iodoalkanes. The first reaction yields phosphorus triiodide, which reacts with the alcohol.

$$P_4 + 6I_2 \rightarrow 4PI_3$$

$$3CH_3CH_2OH + PI_3 \rightarrow 3CH_3CH_2I + H_3PO_3$$

Oxidation with potassium dichromate(VI) in sulfuric acid

In general, balanced equations are not written for these oxidation reactions. The oxidising agent (potassium dichromate(VI) and dilute sulfuric acid) is represented by [O].

Primary alcohols react to give an aldehyde, which, if not removed from the reaction system, oxidises further to yield a carboxylic acid.

$$CH_3CH_2OH + [O] \rightarrow CH_3CHO + H_2O$$

$$CH_3CHO + [O] \rightarrow CH_3COOH$$

The oxidising agent changes colour from orange to green.

Secondary alcohols react to give ketones. The ketones are not oxidised further under these conditions, so the mixture can be heated under reflux.

$$CH_3CH(OH)CH_3 + [O] \rightarrow CH_3COCH_3 + H_2O$$

Tertiary alcohols do not react under these conditions.

Preparation of ethanal by oxidation of ethanol

This standard preparation produces ethanal in aqueous solution. It is invariably contaminated with a small amount of ethanoic acid. The following table lists the steps in the preparation.

Step	Reason
(1) Place 50 cm^3 of water in a 500 cm^3 round-bottomed flask and add slowly, with shaking, 17 cm^3 of concentrated sulfuric acid. Add some anti-bumping granules.	**Why must the sulfuric acid be added slowly and with shaking?** The reaction of sulfuric acid with water is dangerously exothermic. Sulfuric acid is almost twice as dense as water, so if two layers are allowed to form and are then mixed it is possible to generate steam, which could spray the acid about.
	What are anti-bumping granules and what is their purpose? The granules are silica. They prevent the sudden formation of large gas bubbles that lead to 'bumping' (succussion).

Step	Reason
(2) Assemble the flask containing the sulfuric acid into a distillation apparatus. The still head carries a tap funnel instead of a thermometer. The receiving flask should be surrounded by an ice–water bath.	**Why is the receiver surrounded by ice–water?** The boiling temperature of ethanal is 21°C, so the cooling reduces evaporation of the ethanal.
(3) Dissolve 50 g of sodium dichromate(VI) in 50 cm³ of water contained in a small beaker and add 40 cm³ of ethanol. Stir thoroughly and place this mixture in the tap funnel.	**Why isn't the ethanol oxidised by this mixture?** The oxidation requires hydrogen ions, but the mixture as made is not acidic. $Cr_2O_7^{2-} + 14H^+ + 6e^- \rightarrow 2Cr^{3+} + 7H_2O$ **Why is sodium dichromate(VI) used rather than the commoner potassium dichromate(VI)?** Sodium dichromate(VI) is more soluble than the potassium salt in ethanol. Sodium dichromate(VI) is not used in volumetric analysis as it is deliquescent (absorbs water), so it cannot be weighed to make a primary standard solution. Potassium dichromate(VI) does not suffer from this disadvantage.
(4) Heat the dilute acid in the flask until it begins to boil gently and then remove the flame and run the alcohol/dichromate(VI) solution slowly into the flask. As soon as the solution enters the hot acid in the flask, a vigorous reaction occurs and a mixture of ethanal and water containing a little ethanoic acid distils over. The reaction mixture becomes green. The addition of the alcohol/dichromate(VI) mixture should take about 20 minutes. Towards the end of this time, replace the flame under the distilling flask to maintain gentle boiling.	**Why is the alcohol/dichromate(VI) mixture added slowly to the hot acid?** Rapid addition would lead to a large amount of oxidising agent in the reaction mixture and therefore significant oxidation of the ethanal produced to ethanoic acid. Excess of the oxidising agent must be avoided. **Why does the distillate always contain some ethanoic acid?** It is not possible to prevent oxidation of ethanal completely, since it is easily oxidised. **Why does the mixture turn green?** The orange dichromate(VI) ion is reduced to hexaquachromium(III) in this reaction. $[Cr(H_2O)_6]^{3+}$ is green.
(5) When the addition of the alcohol/dichromate(VI) mixture is complete, a moderately concentrated aqueous solution of ethanal collects in the receiver.	**Suggest two reasons why the industrial manufacture of ethanal employs vapour-phase oxidation of ethanol using air over a heated silver catalyst.** **(1)** The use of an expensive reagent like sodium dichromate(VI) would be prohibitively expensive on an industrial scale unless the product was of high value. **(2)** An aqueous solution of ethanal would be formed and this would have to be separated, involving further expense. Vapour-phase oxidation gives the pure aldehyde. Industrial processes are seldom a simple scaling-up of laboratory reactions.

If the ethanol and acidic dichromate(VI) mixture is heated under reflux, the ethanal initially formed is completely oxidised to ethanoic acid. This can then be distilled out of the product mixture.

Halogenoalkanes

Halogenoalkanes are compounds in which one or more of the hydrogen atoms of an alkane have been substituted by halogen atoms. The simplest have the general formula $C_nH_{2n+1}X$ where X is Cl, Br or I. The properties of fluoroalkanes differ from those of the other halogenoalkanes, so they are not considered here.

Primary, secondary and tertiary halogenoalkanes have structures analogous to the corresponding alcohols (from which they can be made). The examples in the following table are isomers of C_4H_9Br.

Primary	Secondary	Tertiary
1-bromobutane ($CH_3CH_2CH_2CH_2Br$)	2-bromobutane ($CH_3CH_2CH(Br)CH_3$)	2-bromo-2-methylpropane [$(CH_3)_3CBr$]

Reactions of halogenoalkanes

The reactions are similar for chloro-, bromo- and iodoalkanes. The rates of reaction increase in the order Cl < Br < I.

Reaction with potassium hydroxide in aqueous solution

Halogenoalkanes heated under reflux with aqueous KOH (or NaOH) give mainly the alcohol in a nucleophilic substitution reaction:

$$CH_3CH_2CH_2CH_2Br + KOH \rightarrow CH_3CH_2CH_2CH_2OH + KBr$$

Some ethanol is often added to improve the miscibility of the reagents. There is always some elimination (see below) at the same time.

Reaction with potassium hydroxide in ethanolic solution

With ethanolic KOH, halogenoalkanes eliminate HX to give an alkene. There is always some substitution at the same time.

$$CH_3CH_2CH_2CH_2Br + KOH \rightarrow CH_3CH_2CH=CH_2 + KBr + H_2O$$

Identifying the halogen atom in a halogenoalkane

(1) Heat the halogenoalkane with sodium hydroxide solution to hydrolyse the halogenoalkane to the alcohol and sodium halide:

$$CH_3Br + NaOH \rightarrow CH_3OH + Na^+ + Br^-$$

(2) Acidify the solution with nitric acid (test with litmus).

(3) Add silver nitrate solution.

A white precipitate soluble in dilute ammonia indicates the formation of a chloride; a cream precipitate soluble in concentrated ammonia indicates a bromide; and a yellow precipitate insoluble in ammonia indicates an iodide. If chloro, bromo and iodo compounds having the same carbon chain are treated under identical conditions with aqueous silver nitrate, the precipitate of silver halide appears first with the iodo compound, followed by the bromo and then the chloro. The carbon–iodine bond is the weakest of the three, as it is the longest and has the lowest polarity.

Apart from the initial hydrolysis, these reactions are the standard tests for halide ions.

Reaction with ammonia

Halogenoalkanes heated in a sealed tube with concentrated ammonia in ethanol give a mixture of amines. The primary amine (RNH_2) can be made the major product by using excess ammonia. The reaction is a nucleophilic substitution:

$$CH_3CH_2CH_2CH_2Br + 2NH_3 \rightarrow CH_3CH_2CH_2CH_2NH_2 + NH_4Br$$

A mixture is obtained because the amine produced is also a nucleophile and attacks any unchanged halogenoalkane.

Preparation of 1-bromobutane

1-bromobutane can be made from butan-1-ol and sodium bromide in 50% sulfuric acid. Below is a standard preparation together with the reasons for the various steps involved.

Step	Reason
(1) Place 30 cm³ of water, 35 g of powdered sodium bromide and 25 cm³ of butan-1-ol in a 250 cm³ round-bottomed flask. Fit a tap funnel to the flask via a still head.	
(2) Place 25 cm³ of concentrated sulfuric acid in the tap funnel and allow the acid to fall drop by drop into the flask, keeping the contents well shaken and cooling occasionally in an ice–water bath.	**Why is the sulfuric acid added slowly? Why is cooling and shaking needed?** Sulfuric acid when diluted with water gives out a great deal of heat, enough sometimes to raise steam, which would cause dangerous splashing. Hot 50% sulfuric acid (produced in the flask) causes significant oxidation of the sodium bromide to bromine, which is useless in this experiment. The yield of 1-bromobutane could therefore be reduced.

Step	Reason
(3) When the addition is complete, replace the tap funnel and still head with a reflux water condenser and gently boil the mixture over a sand bath for about 45 minutes, occasionally shaking the flask gently.	**Why is a sand bath used for heating?** The sand spreads the heating uniformly over the base of the flask. This reduces the likelihood of cracking and of unwanted side reactions (e.g. excessive oxidation either of bromide ions to bromine or of the alcohol to carbon) owing to hot spots. **Why is the mixture heated for 45 minutes?** Most organic reactions are slow because of the need to break strong covalent bonds — the activation energy for the reaction is high.
(4) Remove the reflux condenser and rearrange the apparatus for distillation. Distil off the crude 1-bromobutane (about 30 cm^3).	**Why is the mixture distilled at this stage?** The liquid 1-bromobutane is removed from the involatile sodium salts (mostly sodium hydrogen sulfate at the end of the reaction) and the much less volatile sulfuric acid. The 1-bromobutane is contaminated with water, unchanged butan-1-ol and some sulfuric acid carried over as tiny droplets during the distillation.
(5) Shake the distillate with water in a separating funnel and run off the lower layer of 1-bromobutane. Reject the aqueous layer.	**What does shaking with water achieve?** Water removes sulfuric acid and some of the butan-1-ol. **How do you decide which layer is to be kept?** On the basis of density: 1-bromobutane has a density of 1.28 g cm^{-3}.
(6) Return the 1-bromobutane to the funnel, add about half its volume of concentrated hydrochloric acid and shake. Run off and discard the lower layer of acid.	**Why is concentrated hydrochloric acid added?** The acid protonates the butan-1-ol, giving an ionic species that is more soluble in water than the alcohol itself: $CH_3CH_2CH_2CH_2OH + H^+ \rightarrow CH_3CH_2CH_2CH_2OH_2^+$
(7) Shake the 1-bromobutane cautiously with dilute sodium carbonate solution, carefully releasing the pressure at intervals.	**Why is the mixture shaken with sodium carbonate solution?** This removes hydrochloric acid dissolved in the 1-bromobutane: $Na_2CO_3 + 2HCl \rightarrow 2NaCl + CO_2 + H_2O$ **Why must the pressure be periodically released?** To avoid the stopper being pushed out and product being lost and sprayed all over you. The pressure is due to liberated carbon dioxide.
(8) Run off the lower layer of 1-bromobutane and add some granular anhydrous calcium chloride. Swirl the mixture until the liquid is clear.	**What is the function of the calcium chloride?** Calcium chloride is a drying agent. **What if the mixture isn't clear?** This means that it is not dry. The cloudiness is caused by tiny droplets of water in the 1-bromobutane.
(9) Filter the 1-bromobutane into a clean, dry flask and distil it, collecting the fraction boiling between 99 and 102°C.	**What is the significance of the temperatures quoted?** 1-bromobutane has a boiling temperature of 101.5°C, so the range is narrow enough to ensure that this is the distillate.

A similar reaction with sodium chloride can be used to make 1-chlorobutane. To make 1-iodobutane requires red phosphorus and iodine instead of sulfuric acid and sodium iodide. Sodium iodide is easily oxidised by sulfuric acid and the yield of the organic product would be low.

Uses of halogenoalkanes

- As reactive intermediates in organic synthesis — the halogen atom can be replaced by a variety of other functional groups, in turn leading to many different types of compound. The chart below shows products from a primary halogenoalkane. Make your own chart for the secondary and tertiary compounds.

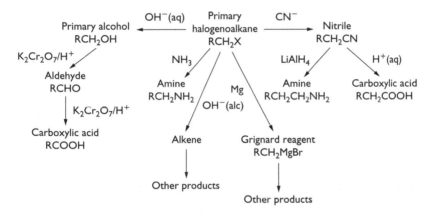

Halogenoalkanes in organic synthesis

- A wide variety of halogenated compounds containing fluoro, chloro and bromo substituents, either alone or in combination, are used as fire retardants or fire extinguishers.
- Chlorofluorocarbons (CFCs) have been used as refrigerants and as aerosol propellants. They are being replaced with hydrofluorocarbons owing to the damage to the ozone layer caused by CFCs.

Mechanisms

A reaction mechanism represents what happens to the electrons as bonds break in the reactants and reform in the products during a reaction. Mechanisms are useful because:

- they enable the classification of an apparently wide variety of reactions into a few categories, showing similarities in the reactions
- they enable the prediction of reaction pathways for new reactions
- if reaction pathways are known in detail, they offer ideas for controlling those pathways in industry or show how they are controlled in nature in biochemical systems

Classifying reactions

The enormous number of chemical reactions fall into one of a few mechanistic categories. Most of the examples given here are organic reactions, but inorganic reaction mechanisms are also well known. Some categories are subdivided into radical, nucleophilic or electrophilic mechanisms.

Addition reactions

Addition reactions occur when two or more molecules combine to give one product:

$$H_2C=CH_2 + H_2 \rightarrow CH_3CH_3$$

$$H_2C=CH_2 + HBr \rightarrow CH_3CH_2Br$$

$$n(H_2C=CH_2) \rightarrow (\ CH_2CH_2-)_n$$

The last example is of **addition polymerisation**.

Elimination reactions

Elimination reactions occur when a molecule loses two or more atoms, turning a single into a double bond. Thus, with KOH in hot ethanol bromoethane eliminates hydrogen bromide:

$$CH_3CH_2Br + KOH \rightarrow H_2C=CH_2 + KBr + H_2O$$

Substitution reactions

Substitution reactions involve the replacement of an atom or group of atoms with another atom or group. Halogenoalkanes undergo **nucleophilic substitution**:

$$CH_3CH_2Br + KOH \rightarrow CH_3CH_2OH + KBr$$

$$CH_3CH_2Br + KCN \rightarrow CH_3CH_2CN + KBr$$

Aromatic compounds (e.g. benzene) undergo **electrophilic substitution**:

$$C_6H_6 + HNO_3 \rightarrow C_6H_5NO_2 + H_2O$$

Oxidation

Oxidation occurs when a species loses electrons. In organic chemistry, oxidation is usually interpreted at AS as a reaction with oxygen or the loss of hydrogen:

$$C_3H_8 + 5O_2 \rightarrow 3CO_2 + 4H_2O$$

$$CH_3CH_2OH \rightarrow CH_3CHO + H_2 \qquad \text{(over hot Cu catalyst)}$$

Oxidising agents such as potassium dichromate(VI) are usually represented in equations as [O], as in the reaction of acidified potassium dichromate(VI) with ethanol:

$$CH_3CH_2OH + [O] \rightarrow CH_3CHO + H_2O$$

Reduction

Reduction occurs when a species gains electrons. In organic chemistry, reduction is usually interpreted at AS as a reaction with hydrogen or the loss of oxygen. Reducing agents other than hydrogen itself are represented as [H].

$$H_2C=CH_2 + H_2 \rightarrow CH_3CH_3$$

$$CH_3CHO + 2[H] \rightarrow CH_3CH_2OH$$

Note that when [O] or [H] are used they must be in the correct number.

Hydrolysis

Hydrolysis is nominally reaction with water, though most hydrolysis reactions are catalysed by acid or alkali (or enzymes). The hydrolysis of esters is typical. Acid hydrolysis gives an equilibrium mixture.

$$CH_3COOCH_2CH_3 + H_2O \rightleftharpoons CH_3COOH + CH_3CH_2OH \text{ (acid hydrolysis)}$$

$$CH_3COOCH_2CH_3 + NaOH \rightarrow CH_3COONa + CH_3CH_2OH \text{ (alkaline hydrolysis)}$$

Polymerisation

Polymerisation is a reaction in which a large number of monomer molecules are joined to make a long-chain molecule with a molar mass in the hundreds of thousands. Alkenes undergo **addition polymerisation**:

$$n(H_2C{=}CH_2) \rightarrow (-CH_2CH_2-)_n$$

Bond breaking: radicals, electrophiles and nucleophiles

Covalent bonds can be broken in two ways. **Homolytic fission** occurs when each fragment retains one of the bonding electrons. The resulting free radicals (also called just 'radicals') have an unpaired electron and are highly reactive. Radicals may be single atoms or groups of atoms. The unpaired electron is shown by a dot on the atom that possesses it:

The movement of a single electron is shown by an arrow with a single barb.

Heterolytic fission occurs when the bond breaks so that both electrons are taken by one of the atoms or groups. The products are then ions:

The positive ion formed may be an **electrophile** — a species that accepts electron pairs in a reaction. Electrophiles in organic reactions include Br^+ (e.g. in reaction of alkenes and bromine) and NO_2^+ (e.g. in reaction of benzene and nitric acid).

The negative ion may be a **nucleophile** — an atom, group or molecule that can donate a lone pair of electrons. Common nucleophiles that are negative ions in organic reactions include Cl^-, Br^-, I^-, OH^- and CH_3O^-. Some nucleophiles are neutral molecules with lone electron pairs — for example, water and ammonia.

Reactions are classified according to the nature of the first attacking reagent. For example, when bromine reacts with ethene, the electrophile $Br^{\delta+}$ formed attacks the ethene, so the reaction is *electrophilic* addition. However, the ethene donates an electron pair from the double bond and is a nucleophile, as is the Br^- that attacks the carbocation in the second step.

Bond polarity and mechanism

The polarity of a bond determines the type of reagent that will attack it.

Alkanes are not polar and have no π-bonds with a high electron density. They are attacked only by radicals — reactive species that can break the strong σ-bonds. The result is **radical substitution**.

The carbon–halogen σ-bond in halogenoalkanes is polar. It has no area of high electron density, so the $\delta+$ carbon atom is attacked by nucleophiles, resulting in **nucleophilic substitution**.

The C=C bond in alkenes and many other compounds is not polar, so it is not attacked by nucleophiles. It contains an accessible cloud of electron density in the π-bond, so it is attacked by electrophiles such as Br_2 or HBr. The result is **electrophilic addition**.

The C=O bond in carbonyl compounds is polar $^{\delta+}C=O^{\delta-}$, because oxygen is more electronegative than carbon. Although the C=O bond contains a π-bond, the more positive carbon atom is attacked by nucleophiles such as cyanide ions or by AlH_4^- ions from $LiAlH_4$. This compound can be used to reduce the C=O group specifically where the C=C group is also present but is not reduced. Thus the compound $CH_3CH=CHCHO$ when treated with $LiAlH_4$ gives $CH_3CH=CHCH_2OH$.

Drawing mechanisms

A drawing of a reaction mechanism is a static description of a dynamic process. By means of curly arrows, which show how electrons move, you are depicting the dynamics of how the electrons change their positions as the reaction proceeds. It is essential to have a picture in your mind of this dynamic process — do not simply regard the mechanism as a series of drawings, but rather as the starting and finishing frames of an animation.

Movement of a single electron is represented by arrows having a single barb. Movement of a pair of electrons is shown by an arrow with a twin barb. Arrows should be drawn with some care because you are showing where electrons start from and where they end up.

> If you draw a mechanism with an intermediate (e.g. the carbocation in an S_N1 mechanism for the reaction of a halogenoalkane with hydroxide ions; see p. 57), which therefore has two steps, it is good practice to draw the intermediate again for the second step (see p. 57) rather than cascading the two steps into one.

Free-radical substitution

An example of a free-radical substitution is the reaction of methane and chlorine in ultraviolet light. The overall reaction is:

$$CH_4 + Cl_2 \rightarrow CH_3Cl + HCl$$

In the **initiation** step, the chlorine undergoes homolytic fission:

$$Cl \overset{\frown}{} Cl \longrightarrow 2Cl \bullet$$

Propagation steps produce a radical for every one that is used:

$$H_3C \overset{\frown}{} H \quad \bullet Cl \longrightarrow \bullet CH_3 + HCl$$

$$Cl \overset{\frown}{} Cl \quad \bullet CH_3 \longrightarrow \bullet Cl + ClCH_3$$

Termination steps do not produce radicals and are responsible for bringing the reaction to an end:

$$\bullet CH_3 \quad \bullet CH_3 \longrightarrow CH_3CH_3$$

$$\bullet CH_3 \quad \bullet Cl \longrightarrow CH_3Cl$$

$$2Cl \bullet \longrightarrow Cl_2$$

Electrophilic addition

Alkenes react by electrophilic addition. The reaction of an alkene with bromine generates the electrophilic $Br^{\delta+}$ by the electrons in the double bond repelling electrons in the Br–Br bond.

In the case of reactions with HBr, the bond is already polarised ($^{\delta+}H–Br^{\delta-}$), so the hydrogen end is electrophilic.

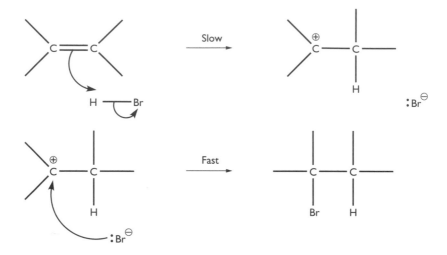

The electrophilic addition of HBr to propene gives, as the major product, 2-bromo-propane. The mechanism involves an intermediate carbocation; the reaction mechanism goes mostly via the most stable carbocation. This is more easily formed than the carbocation that leads to 1-bromopropane.

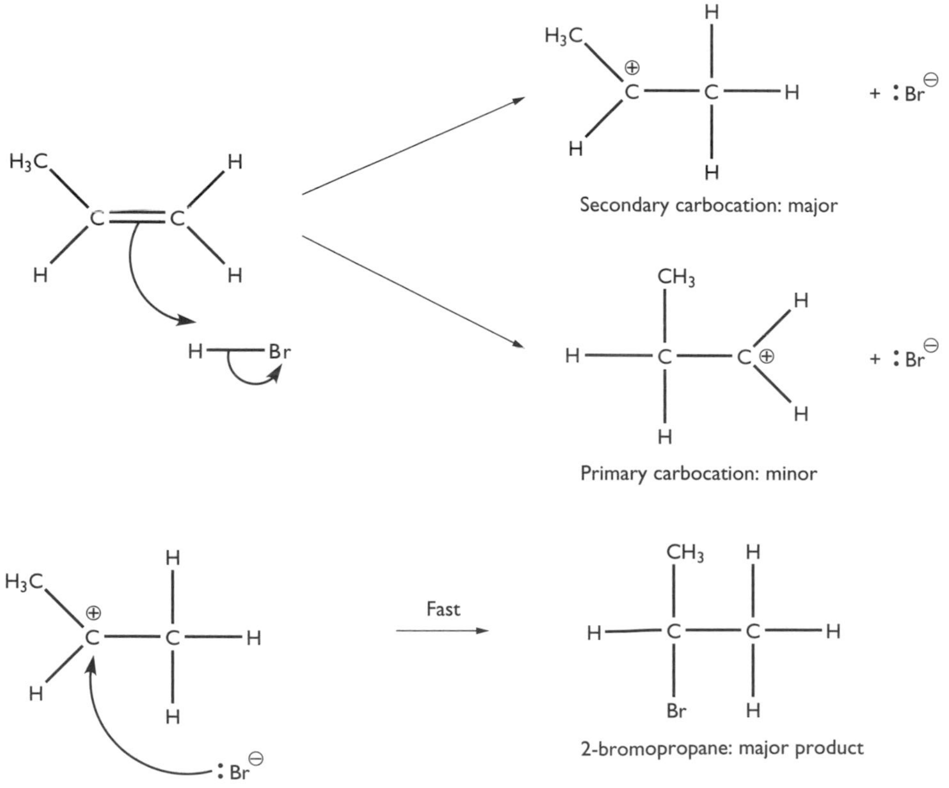

Secondary carbocation: major

Primary carbocation: minor

2-bromopropane: major product

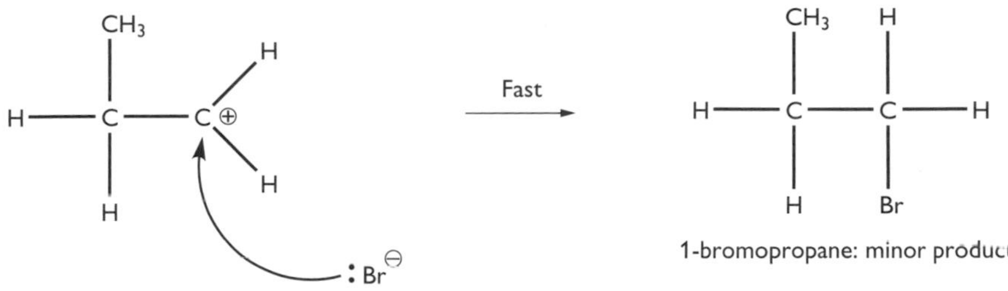

1-bromopropane: minor product

Nucleophilic substitution

There are two types of nucleophilic substitution when hydroxide ions attack a halogenoalkane.

S_N1

The S_N1 reaction (substitution nucleophilic unimolecular) occurs with tertiary halogenoalkanes, such as 2-bromo-2-methylpropane. These compounds are able to generate tertiary carbocations, which are formed more easily than primary or secondary carbocations. The mechanism has two steps. The first step is the rate-determining (slow) step. If the starting halogenoalkane is chiral (see Unit 4), the product mixture has equal amounts of the two optical isomers of the product and so is not optically active.

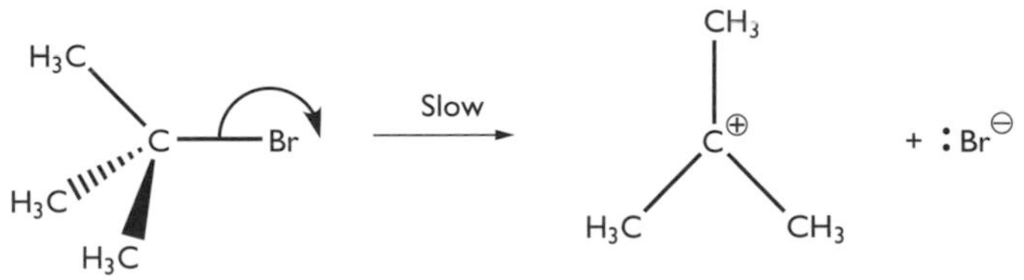

Trigonal planar carbocation

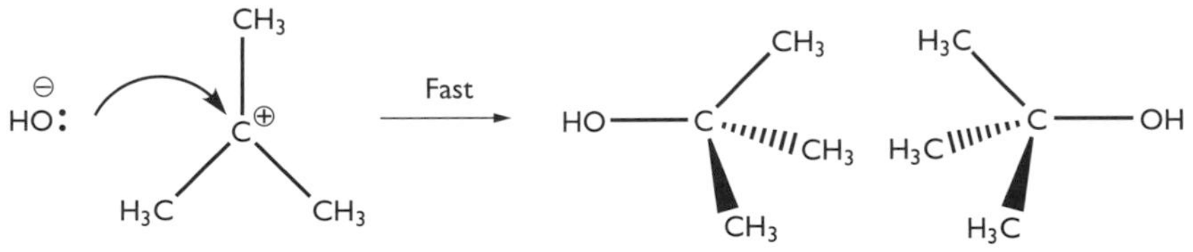

Equal attack from both sides of the planar carbocation

Equal amount of each isomer, but detectable only if the starting halogenoalkane is chiral. The product mixture is then not optically active.

S_N2

The S_N2 reaction (substitution nucleophilic bimolecular) occurs with primary halogenoalkanes, such as bromoethane. Heterolytic fission of the carbon–halogen bond (as in the S_N1 reaction above) would yield a primary carbocation, so the energetically more favoured mechanism has only one step.

A chiral halogenoalkane gives an alcohol in which the arrangement of the substituent groups is inverted (see Unit 4).

In synthetic processes where the stereochemistry of the product is critical (e.g. in the manufacture of pharmaceuticals), a careful mechanistic analysis of the intermediate stages is necessary for the overall synthetic pathway to be successful.

Stratospheric chemistry

The chemistry of the upper atmosphere and ozone depletion is complex and there are many uncertainties. It involves many different reactions that occur in the gas phase and on the surface of clouds and that depend on the season.

The ozone layer occurs at about 20 km above the Earth's surface. It absorbs ultra-violet (UV) light and protects life from excessive amounts of this damaging radiation. In the mid-1980s, data from the British Antarctic Survey showed that the ozone layer over the Antarctic became rapidly depleted at the start of spring. Extensive depletion of ozone is undesirable because of the increased UV that reaches the Earth's surface. The principal materials that destroy the ozone layer are chlorofluorocarbons (CFCs, once used widely in refrigerants and aerosols) and nitrogen oxides emitted by jet engines. Other volatile chlorine- and bromine-containing compounds also break down ozone. (Note that the problems with the ozone 'hole' are not linked to global warming, even though the compounds involved do absorb infrared radiation effectively; see p. 61.)

Both production and destruction of ozone occur in the stratosphere. The two processes are in equilibrium about 20 km above the Earth:

Production: $O_2 \rightarrow 2O\bullet$ (requires UV light of wavelength 240–250 nm)

$O\bullet + O_2 \rightarrow O_3$ (slower with increasing altitude)

Destruction: $O_3 \rightarrow O\bullet + O_2$ (UV light of wavelength 300 nm; faster with increasing altitude)

$$O\bullet + O_3 \rightarrow 2O_2$$

The principal destructive agent for ozone is the chlorine radical (Cl•). In the Antarctic winter there is a column of cold air (the polar vortex) within which polar stratospheric clouds form. The clouds act as surfaces upon which a series of reactions occur that convert long-lived chlorine-containing species into chlorine molecules. The chlorine comes originally from CFCs and other chlorine compounds that are stable to oxidation and hydrolysis and so are not broken down lower in the atmosphere. Oxides of nitrogen are also involved. The intermediate compounds that build up over winter include hydrogen chloride (HCl), chlorine nitrate ($ClONO_2$), dichlorine dioxide (Cl_2O_2) and chlorine dioxide (ClO_2). When sunlight returns to the Antarctic in July, the reservoir of chlorine molecules is converted by homolytic fission to chlorine radicals, which catalyse the decomposition of ozone:

$Cl_2 \rightarrow 2Cl\bullet$

$Cl\bullet + O_3 \rightarrow ClO\bullet + O_2$

$ClO\bullet + O\bullet \rightarrow Cl\bullet + O_2$ etc.

Nitrogen oxides form chlorine nitrate and are involved in other reactions that produce HCl and HOCl.

Mass spectra and IR

Mass spectra

The mass spectroscope was invented by Francis W. Aston, who received the Nobel prize in chemistry in 1922 for using his instrument to detect isotopes. The mass spectrometer is the modern equivalent, one type of which works as follows.

- The vapour of an element or a compound is bombarded with high-energy electrons.
- The electrons ionise the element or the molecule to give positive ions; molecules also form positively ionised fragments.
- The positive ions are accelerated by an electric field and then passed through a magnetic field.
- The magnetic field deflects the ions in a circular path whose radius depends on the mass/charge ratio (m/e) of the ion: the more massive the ion, the less its deflection.
- By changing the strength of the magnetic field, ions of a particular mass/charge ratio can be made to strike a detector, which then measures the abundance of

ions with that specific m/e ratio.
- The machine prints out a bar graph, which shows the abundance of each ion plotted against m/e. This is called the **mass spectrum**.
- The technique is sensitive and fragmentation patterns are characteristic, so they can be used to identify compounds.
- The height of the peak from the most abundant ion is scaled to a value of 100.

Propanal and propanone

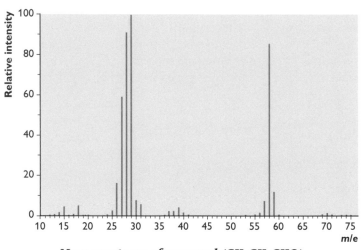

Mass spectrum of propanal (CH₃CH₂CHO)

The most significant peaks in the mass spectrum of propanal are listed below. Remember that all the peaks represent positive ions. Not all are necessarily useful in finding the structure of a substance.
- The peak at $m/e = 58$ is the molecular ion peak arising from $CH_3CH_2CHO^+$. Not all compounds give a molecular ion peak. There is sometimes a small peak one unit higher in mass arising from the presence of ^{13}C in the molecule.
- $CH_3CH_2CO^+$ at $m/e = 57$.
- CHO^+ and $CH_3CH_2^+$ at $m/e = 29$.
- CH_3CH^+ or $CH_2CH_2^+$ at $m/e = 28$.
- CH_3^+ at $m/e = 15$.

Propanone does not have a large peak at $m/e = 29$ because the molecule cannot fragment to give either CHO^+ or $CH_3CH_2^+$.
- The peak at $m/e = 58$ is the molecular ion peak from $CH_3COCH_3^+$.
- There is a small peak at $m/e = 59$ from molecules containing ^{13}C.
- The large peak at $m/e = 43$ is from CH_3CO^+. This is absent in propanal, because the molecule cannot fragment to give this ion.
- CH_3^+ is at $m/e = 15$.

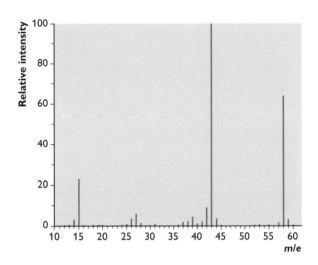

Mass spectrum of propanone (CH₃COCH₃)

IR

Infrared (IR) spectroscopy is a powerful tool for discovering which functional groups are present in a compound. Since most IR spectrometers now have a large database of known compounds installed in memory for comparison, it can also often tell you what a substance is.

Only compounds with polar bonds are able to absorb infrared radiation. As they absorb infrared radiation, the polarity of the molecule changes.

The absorption frequencies of a molecule depend on the type of the bond. The absorptions in the spectra below are from liquid film samples and are all stretching vibrations.

The horizontal scale is calibrated in wavenumbers, which are measured in units of cm^{-1}. This unit of cm^{-1} is used only to measure IR spectra. It refers to the number of wavelengths of the radiation per centimetre. A wavenumber of 3000 cm^{-1} corresponds to a frequency of about 10^{14} Hz.

The spectrum from about 1400 cm^{-1} to 600 cm^{-1} is called the **fingerprint region** and corresponds to bending vibrations of the molecule. Its complexity makes it useful for comparison between an unknown spectrum and a reference database since each compound has a unique fingerprint.

The vertical axis is calibrated as percentage transmittance, so IR absorptions appear as troughs or dips in the spectrum. They are nevertheless sometimes referred to as peaks.

Types of bond stretch

C–H stretch

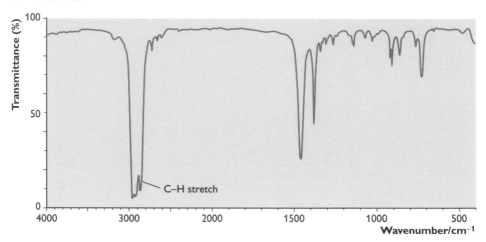

IR spectrum of pentane (CH₃CH₂CH₂CH₂CH₃)

The C–H stretch in pentane occurs at approximately 2800–3000 cm⁻¹. Almost all organic molecules have C–H bonds and have a large absorbance in this region, so it is not useful for identifying substances.

O–H stretch

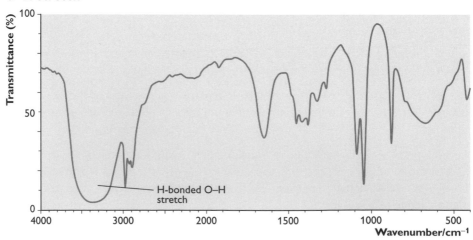

IR spectrum of ethanol (CH₃CH₂OH)

The hydrogen-bonded O–H stretch in ethanol occurs at about 3200–3800 cm⁻¹. The frequency at which the O–H bond vibrates depends on how tightly it is hydrogen-bonded to another molecule. This differs between molecules so the absorption peak is broad. If the ethanol is diluted with a solvent that does not form hydrogen bonds, the peak narrows and moves to about 3400 cm⁻¹.

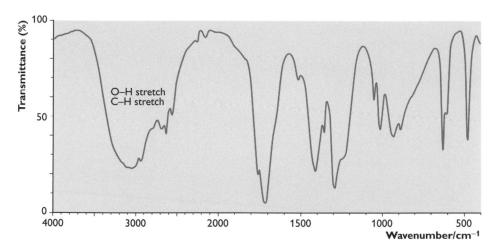

IR spectrum of ethanoic acid (CH₃COOH)

The O–H stretch in carboxylic acids occurs at about 2500–3300 cm⁻¹. The hydroxyl group is hydrogen-bonded so as to form a dimer (two molecules joined together, as shown below). The O–H absorption is broad and lies over the C–H stretch.

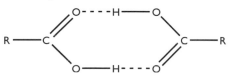

N–H stretch

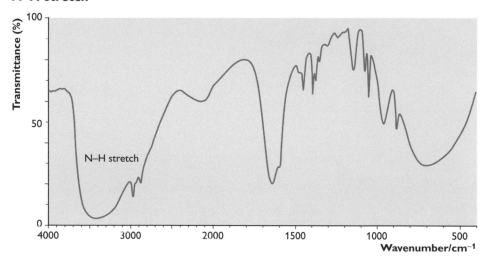

IR spectrum of ethylamine (CH₃CH₂NH₂)

The hydrogen-bonded N–H stretch in amines is broad because of hydrogen bonding.

C=O stretch

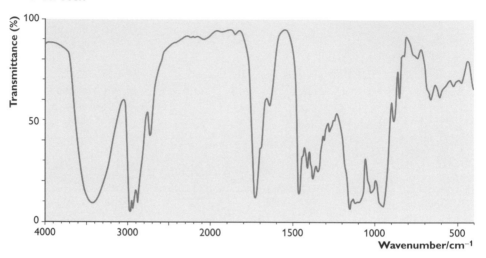

IR spectrum of propanal (CH₃CH₂CHO)

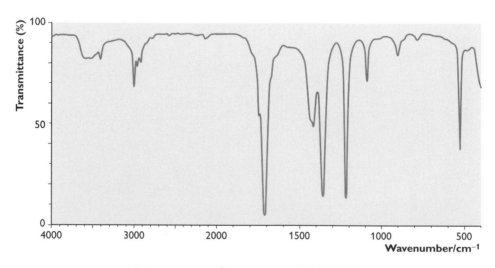

IR spectrum of propanone (CH₃COCH₃)

The C=O stretching vibration in both propanal and propanone is just above 1700 cm⁻¹. Propanal also shows the broad O–H stretch around 3400 cm⁻¹. This is due to the existence of the following equilibrium in the liquid:

$$CH_3CH_2CHO \rightleftharpoons CH_3CH=CHOH$$

C–X stretch (halogenoalkanes)

The C–X stretch in halogenoalkanes occurs at about 800–600 cm⁻¹. This absorption is usually hard to see, as it lies in the complex fingerprint region of the spectrum and it is difficult to determine which absorption is which.

content guidance

Following the progress of a reaction

IR spectroscopy can be used to monitor the progress of a reaction — for example, the oxidation of an alcohol. Primary and secondary alcohols are oxidised by acidified potassium dichromate(VI) to give aldehydes or ketones. This is readily seen in the IR spectrum of the reactant alcohol, which has a broad O–H stretch at 3200–3800 cm^{-1} but no carbonyl absorption around 1700 cm^{-1}. As the reaction progresses, the peak due to the O–H stretch disappears and a new peak appears at around 1700 cm^{-1}, due to the carbonyl group in the product.

Greenhouse gases

If the greenhouse effect did not exist, the Earth would be uninhabitable. It would be extremely hot during the day and extremely cold during the night.

The atmospheric gases that absorb IR radiation are water (H_2O), carbon dioxide (CO_2), methane (CH_4) and nitrogen oxides (NO_x), all of which change their polarity when exposed to IR. Water vapour is the most significant absorber. Methane is about 30 times and nitrogen oxides are about 160 times more absorbing than carbon dioxide, molecule for molecule. However, their overall contribution to the greenhouse effect is complex and is not simply the sum of each factor multiplied by the abundance. It also varies with height in the atmosphere. (Oxygen and nitrogen are not polar molecules, so they do not absorb IR and do not contribute to the greenhouse effect.)

Green chemistry

Most of the chemistry covered in the earlier sections of this guide is well understood, but much of the science involved in the debate on climate change is complex and different models produce many possible outcomes, depending on the initial assumptions. The Earth is also dynamic and has a large degree of inbuilt adaptation to changing circumstances, but the extent to which it can adapt is unclear. However, tackling problems of climate change is only partly scientific, as they are mostly political and social, but chemistry can illuminate the debate.

Sustainable chemistry

The planet's chemical resources are finite. The more people there are, the greater the need to manage the resources carefully and to ensure that they are re-used as much as possible (bearing in mind that recycling also has costs). It is important, when considering alternative materials, to look at all the energy requirements for manufacturing, the implications for disposal and whether the manufacturing processes can be made more efficient and safe. The use of processes with higher atom economies and the development of increasingly efficient catalysts will produce a chemical industry that has a decreasing impact on the environment. It is important to remember, though, that humans cannot exist without affecting their environment — this is built into the thermodynamics of organisms.

Reducing chemical hazards

Some substances are very hazardous — they have the potential to do serious damage to living things. One of the skills that chemistry teaches is the safe handling of hazardous materials, but clearly it is in our interests to reduce the use of such substances wherever possible. In Unit 1 you learned that the risk from a hazardous material can be reduced by using less of it or by using an appropriate containment method. The best method is to avoid the material, but this is not always possible — for example, chlorine is important in maintaining public health and its only alternative for water treatment (ozone) is equally hazardous.

In some cases, there are alternatives. The use of CFCs as refrigerants is gradually being phased out in favour of hydrofluorocarbons, which do not destroy ozone. Chlorinated solvents as degreasing agents are being replaced with materials that do not contain halogens.

Catalysis

A catalyst provides a pathway for a reaction that has a lower activation energy than the uncatalysed pathway. This makes the reaction faster and requires lower temperatures and less energy. Enormous effort is being expended on developing catalysts that are highly specific, so that a process gives only the material required and suppresses side reactions. This leads to higher atom economies.

Ethanoic acid has many uses — for example, in foodstuffs (vinegar), in making emulsion paints, textiles, solvents and pharmaceuticals. Over 6 million tonnes of ethanoic acid is made every year, including 0.5 million tonnes produced in the UK. In the 1960s, ethanoic acid was produced using methanol and carbon monoxide, which were made from natural gas (coal or oil can also be used to make methanol). The original catalyst was a cobalt compound combined with iodine, but later developments used rhodium and iodine. Rhodium is in the same group as cobalt, but the catalyst is more efficient, works at lower temperatures (450–460 K) and pressure (30 atm) and does not generate waste products.

Using high pressures is expensive, not principally because of the cost of the manufacturing plant (which is high for the rhodium process because special materials are needed for construction) but because of the high daily cost of fuel for running the gas compressors. This outweighs the cost of every other aspect of high-pressure operation.

Heating

Fuel costs are usually the second-highest cost (after wages) of most commercial operations. In the chemical industry, fuel can be the biggest cost, so reducing fuel consumption is important. Microwave heating is being used increasingly for small-scale processes, e.g. in the pharmaceutical industry. It is more efficient and controllable than conventional heating but cannot be used for large-scale processes.

Reducing waste

Industrial processes are not generally laboratory processes scaled up. There must be few waste materials in industry, because they are part of the production cost and also because waste disposal is expensive. Oxidation of an alcohol using potassium dichromate and sulfuric acid is often used in the laboratory, as the production of a few grams of chromium(III) salts is not expensive or polluting. However, on an industrial scale, it is more attractive to use catalysts and readily available oxidants (e.g. air) that give only the oxidised product with no waste.

Efficiency in production leads to less pollution. Regulatory bodies exert stringent controls on industry, so that air and water are now less polluted than at any time since the start of the Industrial Revolution.

Global warming

Without global warming the Earth would be uninhabitable. The days would be intolerably hot and the nights intolerably cold. The topical question of the extent to which human activity contributes to climate change is controversial, as the chemistry of the planet is complex and there are many uncertainties in the science.

Climate change has always occurred, as shown by the geological and fossil record over millions of years and by more recent data on the 'mini' ice age in medieval times. However, the question of whether rapid climate change is now occurring as a result of industrial human activity (anthropogenic climate change) is a hot topic. It raises the political, social and economic questions of how much, and what, needs to be done — and how. The *precautionary principle* says that all possible steps must be taken as soon as possible to combat the worst scenario predicted by current models, but this would be too expensive and have profound effects on how we live — and might be unnecessary anyway. There are many climate models for the future, with differing initial assumptions and final outcomes, all based on science that is still poorly understood. There are carbon cycles in the air, on land and in the seas. The way in which they interact and how they might be affected by changing temperatures, for example, is uncertain.

Greenhouse gases

The most plentiful and effective greenhouse gas in the atmosphere is water vapour. Much of the popular discussion of global warming centres on carbon dioxide with occasional reference to nitrogen oxides. Carbon dioxide has almost attained the status of a poisonous gas in the media, so it is worth remembering that every carbon atom in your body, without exception, came from carbon dioxide.

Some atmospheric gases are more effective at absorbing IR than others. Methane is 30 times more absorbing, molecule for molecule, than carbon dioxide is, nitrogen oxides about 150–200 times and CFCs over 20 000 times as absorbing. The ability to contribute to global warming is more complex and is not simply the weighted average

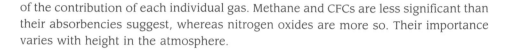

of the contribution of each individual gas. Methane and CFCs are less significant than their absorbencies suggest, whereas nitrogen oxides are more so. Their importance varies with height in the atmosphere.

Carbon neutrality

Carbon footprint

A process that is carbon-neutral makes no net contribution to the amount of carbon dioxide in the atmosphere. The term 'carbon footprint' does not yet have a universally agreed definition. For many people it refers only to the amount of carbon dioxide produced in the atmosphere from the system being considered, but others would also include methane and nitrogen oxides because they are greenhouse gases. There are also difficulties in deciding where the boundaries should be drawn when determining carbon dioxide emissions. For example, consider the manufacture and use of a bicycle, a favourite form of green transport. The following sources of carbon dioxide in its manufacture come from the fuel used:

- to mine and transport the ores for the metal used in its construction
- to smelt the metals and transport them to the factory
- in making the frame, wheels etc.
- to fractionate crude oil and process it to make the tyres, lubricants, brake pads and paints
- directly in the manufacturing process to drive machines and to heat and light the factory
- in disposing of the bicycle at the end of its useful life and recycling its components (as well as carbon dioxide produced from the decomposition of its organic constituents)

In addition, there is the carbon dioxide from the fuel used to manufacture the machine tools in the factory, or even that produced to make the materials to build the factory. Perhaps the carbon dioxide generated by the workers travelling to the factory should also be considered.

The Carbon Trust considers only the carbon dioxide from the input, output and unit processes directly associated with a product, so it would not include travel by workers.

Another definition of carbon footprint is as 'a measure of the total amount of carbon dioxide emission that is directly or indirectly caused by an activity or accumulated over the life stages of a product'. This could include goods or services.

There is no agreement on where the boundaries should be drawn. The problems with defining the carbon footprint are not entirely scientific but are also political and social.

Fuels

Fossil fuels

Fossil fuels have locked up, for several hundred million years, large amounts of carbon dioxide that used to be in the atmosphere. Burning of any fossil fuel (whether coal, petrol, diesel or natural gas) returns this carbon dioxide to the atmosphere. Such fuels are not carbon-neutral, as they increase the amount of atmospheric carbon dioxide that has to be dealt with by natural carbon cycles.

Bioethanol

Bioethanol seems to be a carbon-neutral fuel. Carbon dioxide is absorbed by the plant as it grows and is released again when the fuel is burnt. However, there are serious problems with the large-scale use of biofuels. First, they take up land that could otherwise be used to grow food crops. This has contributed to difficulties with some food supplies in recent years. Second, many biofuel crops such as maize or soya are being grown on cleared rainforest land. This lowers the ability of that land to absorb carbon dioxide, since the crops use less of the gas for photosynthesis than the carbon-rich rainforest that they replaced. Third, fermentation needs energy, so at best the advantages of biofuel are less dramatic than might at first be apparent and in some cases they are harmful. The hidden carbon costs may make the overall saving negligible.

Hydrogen

Hydrogen produces only water when it burns, so it is regarded as an attractive fuel in cities. However, hydrogen as a fuel is far from carbon-neutral, as it has to be manufactured from water and methane. Purified methane and steam are passed over a nickel oxide catalyst at about 1000 K:

$$CH_4(g) + H_2O(g) \rightarrow CO(g) + 3H_2(g) \qquad \Delta H = +210 \text{ kJ mol}^{-1}$$

This process requires energy for heating, both for the purification of the methane and for its reaction with steam. Some energy comes from burning the carbon monoxide that is generated, but the production of hydrogen clearly involves carbon dioxide emissions.

Carbon offsetting and CO_2 control

Whenever a process is invented to reduce carbon dioxide emissions from the burning of fossil fuels, it is important that the contributions from all aspects of the new process are considered. There may not seem to be much carbon dioxide involved in running wind farms, but what about the carbon dioxide produced to make the large concrete plinths on which the windmills stand? The 'carbon payback time' is the number of years taken for savings from avoiding the burning of fossil fuels (via a new process of energy generation) to offset losses in ecosystem clearing of carbon due to reductions in forest cover or loss of other agriculture, or from carbon dioxide produced during the manufacture of a new device. The Meteorological Office gives 300–1500 years as the carbon payback time for biofuel produced from cleared rainforest. This seems to be a bad deal.

There are ways to control carbon dioxide emissions from factories — for example, crushed limestone suspended in water can absorb carbon dioxide from flue gases. There are also suggestions that carbon dioxide could be pumped into old oil or gas wells that are no longer in production, thus burying the gas. Such carbon dioxide is **sequestered**.

Chlorofluorocarbons and stratospheric ozone

The mechanisms for the reactions that cause chlorofluorocarbons to damage the ozone layer are given on p. 58.

The decision to end the use of CFCs was based on good evidence of the mechanisms involved in the chemistry of the stratosphere. The properties of CFCs in the test-tube in a laboratory, where they are stable to hydrolysis and oxidation, are not mimicked in the stratosphere in the presence of large amounts of UV light. CFCs have many properties that make them useful as aerosol propellants and as refrigerants, so the challenge for chemists is to produce other substances that can function just as well but do not have the same side effects. This cannot be done overnight. For example, it would be absurd to ban CFC refrigerants before a suitable replacement has been found: refrigeration marked a major advance in maintaining public health and avoiding food wastage.

Questions
&
Answers

This section contains multiple-choice and structured questions similar to those you can expect to find in Unit Test 2. The questions given here are not balanced in terms of types of question or level of demand — they are not intended to typify real papers, only the sorts of questions that could be asked.

In the examinations, the answers are written on the question paper. Here, the questions are not shown in examination paper format.

The answers given are those that examiners would expect from a grade-A candidate. They are not 'model answers' to be regurgitated without understanding. In answers that require more extended writing, it is usually the ideas that count rather than the form of words used. The principle is that correct and relevant chemistry scores.

Before you start to read and use this section, re-read the material on answering examination questions in the introduction to this guide.

Examiner's comments

Responses to questions may have an examiner's comment, preceded by the icon 🖉. The comments may explain the correct answer, point out common errors made by candidates who produce work of C-grade or lower, or contain additional useful advice.

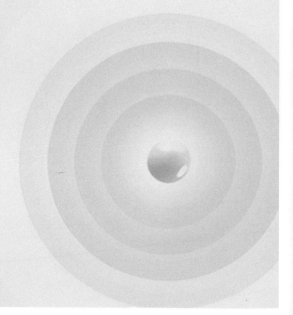

Multiple-choice questions

Each question or incomplete statement is followed by four suggested answers, A, B, C and D. Select the *best* answer in each case. The answers are given, with some commentary, after question 10.

1 Deuterium oxide, D_2O, is water that contains the stable hydrogen isotope ^2H, usually shown as D. When dissolved in deuterium oxide, ethanol undergoes the following reaction:

$$CH_3CH_2OH + D_2O \rightarrow CH_3CH_2OD + HOD$$

The best way of showing that this reaction has occurred is:

A by testing the product with phosphorus pentachloride and not getting steamy fumes

B by determining the mass spectrum of the mixture and showing peaks at *m/e* values of 20 and 46

C by fractionally distilling the mixture and determining the amount of CH_3CH_2OD using a Geiger counter

D by determining the mass spectrum of the mixture and showing peaks at *m/e* values of 19 and 47

2 Which of A–D represents most accurately the hydrogen bonding that occurs between ethanol and water?

A

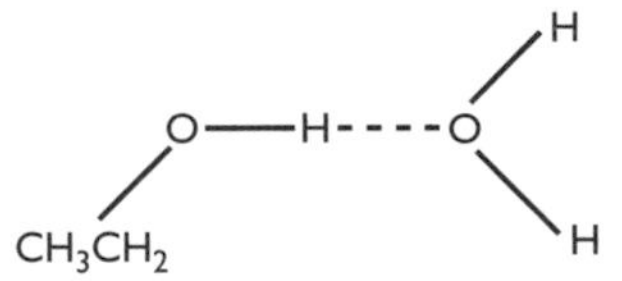

B

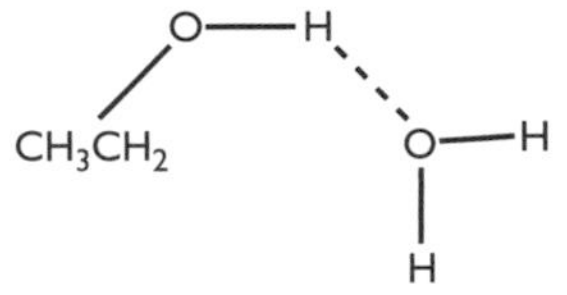

C

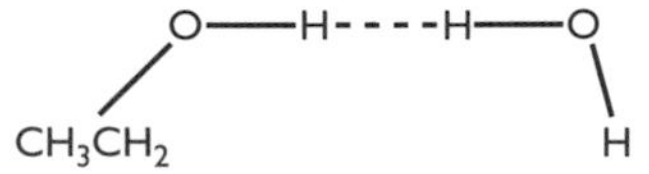

D

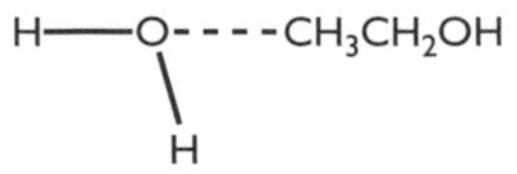

3 A catalyst increases the rate of a chemical reaction by:

A lowering the activation energy for the reaction

B giving the reacting molecules more energy so more of them collide successfully per unit time

C providing an alternative pathway with an activation energy that is lower than that of the uncatalysed reaction

D providing a surface on which the reactants can combine

4 The shape of an ammonia molecule is:
  A trigonal planar with three electron pairs around the nitrogen atom
  B tetrahedral with four electron pairs around the nitrogen atom
  C pyramidal with four electron pairs around the nitrogen atom
  D square planar with four electron pairs around the nitrogen atom

5 Radium occurs below barium in group 2. On the basis of the trends in the chemistry of group 2, which of the following statements is true?
  A radium sulfate is more soluble in water than magnesium sulfate
  B radium hydroxide is more soluble in water than calcium hydroxide
  C radium carbonate is more easily decomposed by heating than magnesium carbonate
  D radium carbonate is very soluble in water

6 A nucleophile is a species that:
  A donates an electron pair
  B accepts an electron pair
  C must be negatively charged
  D has a single unpaired electron

7 When iron(II) sulfate is oxidised by potassium manganate(VII), the iron(II) ions are oxidised to iron(III), and the manganate(VII) ions are reduced to Mn(II).

  50.0 cm^3 of acidified iron(II) sulfate solution of concentration 0.0200 mol dm^{-3} is completely oxidised by x cm^3 of potassium manganate(VII) of concentration 0.0200 mol dm^{-3}. What is the value of x?
  A 10.0
  B 25.0
  C 50.0
  D 200.0

8 Which anions that contain chlorine are produced when chlorine gas is passed into cold, dilute sodium hydroxide solution?
  A Cl^-, ClO_3^-
  B OCl^-, Cl^-
  C ClO_3^-, ClO_4^-, Cl^-
  D ClO_3^-, OCl^-

9 An alcohol with the formula C_3H_8O is heated under reflux with an acidified solution of potassium dichromate. The organic product has the formula $C_3H_6O_2$. Which of the following statements is true?
  A the alcohol is a primary alcohol and it has been oxidised to an aldehyde
  B the alcohol is a secondary alcohol and it has been oxidised to a ketone
  C the alcohol is a primary alcohol and it has been oxidised to a carboxylic acid
  D the alcohol is a tertiary alcohol

10 It is suggested that IR spectroscopy can be used to show that an alcohol has been oxidised to a ketone by comparing the IR spectra of the alcohol and its oxidation product. Which of the following statements is true?

A the alcohol has been oxidised to a ketone if the broad **O–H** absorption at about 3300 cm^{-1} in the alcohol spectrum is replaced by a narrow **C=O** absorption at about 1700 cm^{-1} in the product spectrum

B the alcohol has been oxidised to a ketone if the narrow **O–H** absorption at 3300 cm^{-1} in the alcohol spectrum is replaced by a broad **C=O** absorption at about 1700 cm^{-1} in the product spectrum

C you cannot tell whether the alcohol has been oxidised to a ketone because these compounds do not absorb in the **IR** part of the spectrum

D you cannot tell whether the alcohol has been oxidised to a ketone because both ketones and aldehydes give an **IR** absorption at around 1700 cm^{-1}

■ ■ ■

Answers

1 D

The two peaks mentioned are due to the presence of CH_3CH_2OD and HOD, which would not have been formed in the absence of any reaction. The suggestion for the use of a Geiger counter (option **C**) might be chosen by students who think that the word 'isotope' is synonymous with 'radioactive'.

2 A

In a hydrogen bond, the hydrogen atom forms a bridge between two electronegative atoms; in this case, both are oxygen. However, the hydrogen bond is also linear: O–H-----O is, therefore, a more accurate representation than the diagram shown in option **B**.

3 C

The essential feature of a catalyst is that it changes the mechanism of a reaction to one that has lower activation energy than the uncatalysed reaction. The activation energy for a particular reaction mechanism is intrinsic to that mechanism and cannot be changed. Option **D** is often true, but does not apply to all catalysts.

4 C

The nitrogen atom in ammonia has three bond pairs and one lone pair. The basis of its shape is the (nearly) tetrahedral arrangement of these electron pairs, but the name of its shape is defined by the atom centres and is pyramidal.

5 B

The solubility of the hydroxides of group 2 metals in water increases with increasing atomic mass of the metal (i.e. down the group). All the other suggestions are opposite to the actual trends in the group.

6 A

📝 All nucleophiles donate an electron pair, but they do not have to be negatively charged. Both water and ammonia are nucleophilic reagents.

7 A

📝 Oxidation of Fe^{2+} to Fe^{3+} is a one-electron change; reduction of MnO_4^- to Mn^{2+} is a five-electron change. The concentrations of the solutions are the same, so the volume of manganate(VII) solution required is one-fifth that of the iron(II) solution.

8 B

📝 The ionic equation for the reaction is:

$$Cl_2 + 2OH^- \rightarrow Cl^- + OCl^- + H_2O$$

The chlorine has disproportionated. The answer would be option **A** if the sodium hydroxide solution had been hot, since OCl^- ions disproportionate on heating to chloride ions and chlorate(v) ions.

9 C

📝 Since the only product options for alcohol oxidation are aldehydes, ketones (both with only one oxygen atom) and carboxylic acids, the latter must be the product.

10 D

📝 The O–H absorption is broad and the C=O absorption is narrow, so option A is true, but the point is that aldehydes and ketones cannot be distinguished by their C=O peaks.

Structured questions

Question 11

The reaction between sulfur dioxide and oxygen in a closed system is in dynamic equilibrium:

$$2SO_2(g) + O_2(g) \rightleftharpoons 2SO_3(g) \qquad \Delta H = -196 \text{ kJ mol}^{-1}$$

(a) Explain what is meant by *dynamic equilibrium*. (2 marks)

(b) State the effect on the position of equilibrium of this reaction of:
  (i) increasing the temperature
  (ii) increasing the pressure (2 marks)

(c) This reaction is the first step in the industrial production of sulfuric acid. A temperature of 450°C, a pressure of 2 atm and a catalyst are used. Justify the use of these conditions:
  (i) temperature of 450°C (2 marks)
  (ii) pressure of 2 atm (2 marks)
  (iii) catalyst (1 mark)

(d) Name the catalyst used. (1 mark)

(e) Explain why an industrial manufacturing process such as this cannot be in equilibrium. (2 marks)

Total: 12 marks

■ ■ ■

Answers to question 11

(a) The forward and backward reactions are occurring at the same rate ✓, so there is no net change in composition ✓.

🖉 The first mark is for *dynamic* and the second for *equilibrium*.

(b) (i) Moves to the left *or* concentration of products decreases ✓
  (ii) Moves to right *or* concentration of products increases ✓

🖉 The equilibrium mixture does not move in a physical sense. The terms are used as shorthand — 'moves to the right' means that the concentrations of the substances on the right-hand side of the equation (the products, conventionally) increase. The effect of an increase in pressure on the rate of a heterogeneously catalysed reaction is negligible. The reactants have a strong affinity for the surface of the catalyst. At all pressures (other than extremely low), the catalyst surface, which is where the reaction occurs, is saturated with the reactants and so increasing the pressure makes no difference. This point is commonly misunderstood.

(c) (i) The temperature needs to be high to give an acceptable rate but low to achieve an acceptable yield ✓; the temperature used is therefore a compromise *or* the temperature is the optimum for the catalyst used ✓.

🕑 When the equilibrium temperature is increased, the equilibrium composition changes so that heat energy is absorbed, but this does not bring the temperature down again. The changes are to the *equilibrium* conditions — the alteration in composition has no effect on the conditions, which are externally imposed.

(ii) High pressure increases the yield but is not necessary since an acceptable yield is obtainable at low pressures; *or* you need a high enough pressure to drive the gases through the catalyst bed(s) ✓; an increase in pressure would increase yield but this would not offset the increase in costs ✓.

🕑 The yield is high because, in practice, four catalyst beds are used at different temperatures and these are efficient enough at low pressure. High-pressure sulfur trioxide is extremely corrosive. The use of high pressure would also cause the sulfur dioxide to liquefy in those parts of the plant that are at low temperature. The biggest cost in using high pressure is the cost of fuel for compressing the gases.

(iii) Increases reaction rate ✓

🕑 The implication from a cost–yield viewpoint is that the rate becomes acceptable at a lower temperature than would otherwise be the case.

(d) Vanadium(V) oxide ✓ *or* vanadium pentoxide

🕑 Platinum can be used in the laboratory, but it is not used industrially, so it is not an acceptable answer.

(e) Products are being removed to be sold ✓ so the system is not closed and the reverse reaction cannot occur quickly enough to achieve equilibrium ✓.

🕑 A closed system is one in which no exchange of matter is possible with the surroundings. Reactions in test tubes are in open systems since gases can be evolved. Equilibrium ideas can be used to approximately predict optimum conditions in industrial processes, but these processes cannot be in closed systems.

Question 12

The rate of a chemical reaction increases as the temperature is increased.

(a) Draw a diagram of the Maxwell–Boltzmann distribution of molecular energies at a temperature T_1 and at a higher temperature T_2. (3 marks)

(b) Mark *on your diagram* in a suitable position the activation energy and use this to explain why the rate of a chemical reaction increases with increasing temperature. (4 marks)

Total: 7 marks

■ ■ ■

Answers to question 12

(a)

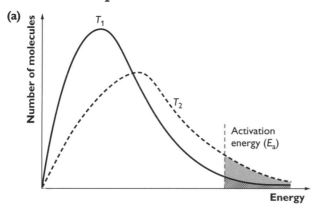

[e] Correctly labelled axes ✓; curve for T_1 skewed and correct shape within reason (starts at the origin and does not intercept the x-axis at high energy) ✓; curve for T_2 with peak to the right and lower ✓.

The Maxwell–Boltzmann curve has a well-defined shape and you are expected to be able to reproduce it accurately. Common errors include drawing a curve that is symmetrical or a curve that either starts part way up the y-axis or intersects the x-axis. The Maxwell–Boltzmann curve is not like a normal distribution (bell) curve.

(b) E_a shown on the diagram well to the right of the peak ✓; area under the curve for energies above E_a is higher for T_2 than for T_1 ✓, so there is a higher proportion of successful ✓ collisions, since collision energy > E_a for these molecules ✓.

[e] The idea of collision energy is important, as is the ability to see kinetics in terms of particle interactions. If you do not refer to the diagram, you can earn a maximum of only 3 marks.

The higher proportion of successful collisions is an important, but subtle, point. It is not true to say, as many candidates do, that the number of collisions increases. Overall, the number of collisions leading to reaction for the same amount of product is the same. It is the time during which these collisions occur that matters. You could also say that the number of successful collisions per unit time increases.

Remember that your script is scanned before marking, so different colours cannot be distinguished and red is not picked up. When you shade the areas under the curves, use different types of shading or hatching, not colours.

Question 13

Consider the following series of reactions, and then answer the questions that follow:

$$C_4H_8 \xrightarrow[\text{ethanolic KOH}]{\text{HBr}} C_4H_9Br \xrightarrow{\text{NaOH(aq)}} C_4H_9OH$$
$$\quad A \qquad\qquad\qquad B \qquad\qquad\qquad C$$

(a) (i) Compound A is an alkene that has two geometric isomers. Draw their
     structural formulae.   (2 marks)

  (ii) State the *two* features of the molecule that makes this isomerism
     possible.   (2 marks)

  (iii) Stating what you would *see*, give a simple chemical test for the functional
     group present in compound A.   (2 marks)

  (iv) What *type* of reaction is the conversion of compound A to compound B?   (1 mark)

  (v) Draw the mechanism for the reaction of compound A with hydrogen
     bromide. In your mechanism, the alkene (A) can be represented simply as:

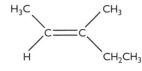

     (3 marks)

  (vi) What type of reaction is the conversion of compound B to compound A?   (1 mark)

(b) (i) The reaction of compound B to give compound C is a nucleophilic
     substitution. What is meant by the term nucleophile?   (1 mark)

  (ii) Give the full structural formula of compound C.   (1 mark)

  (iii) Give a simple chemical test for the functional group in compound C.
     Describe what you would see as a result of this test.   (2 marks)

  (iv) Give the full structural formula for the compound obtained by heating
     compound C with acidified potassium dichromate solution.   (2 marks)

(c) The alkene 3-methylpent-2-ene has two geometric isomers. Giving your
   reasons, state whether the isomer shown below is the *E*-isomer or the *Z*-isomer.

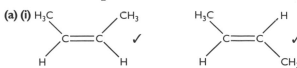

     (2 marks)
     Total: 19 marks

■ ■ ■

Answers to question 13

(a) (i)
H₃C―C=C―CH₃ with H, H ✓   H₃C―C=C―H with H, CH₃ ✓

Each correct structure scores 1 mark.

(ii) Restricted (or no) rotation about the C=C bond ✓ and the two groups on a given carbon in the double bond are different ✓.

📝 The requirement that the groups on a given carbon must be different is often missed.

(iii) Bromine water ✓ is decolorised from yellow (or orange) to colourless ✓.

📝 Note the difference between colourless (no colour) and clear, which means see-through and can be any colour. Confusion between the two is common. The fail-safe answer to all questions involving colour changes is to give the starting and finishing colours.

(iv) Electrophilic addition ✓

(v)

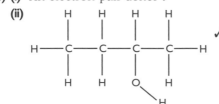

📝 The marks are for the two arrows initially ✓, for the structure of the carbocation intermediate ✓, and for the arrow in the second step ✓ to give the product.

The arrows show the movement of electron pairs and it is important to show their positions accurately. It is not essential to show the lone pair on the bromide ion, but it is better to include it because the lone pair makes the bromide a nucleophile. Similarly the labels 'slow' and 'fast' do not have to be added, but the statements are true so why not include them? Candidates seldom do — but it is knowledge of such details that indicates A-grade performance.

(vi) Elimination ✓

(b) (i) An electron-pair donor ✓

(ii)

📝 The use of 'full' in the question suggests that the displayed formula is the one to use — and it is the one to use if in any doubt. The O–H bond in the hydroxyl group is shown explicitly. Using –OH might be acceptable, but the above approach is fail-safe.

(iii) Add phosphorus pentachloride ✓; steamy (acidic) fumes ✓

e 'Acid fumes' alone would not score because you cannot see that the fumes are acidic; nor are the fumes white, which is a common error.

(iv)

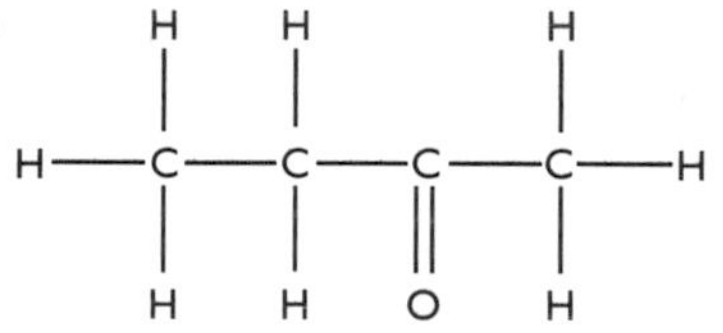

✓✓

e There is 1 mark for the structure being a ketone and 1 mark for it being the correct ketone.

(c) In the structure:

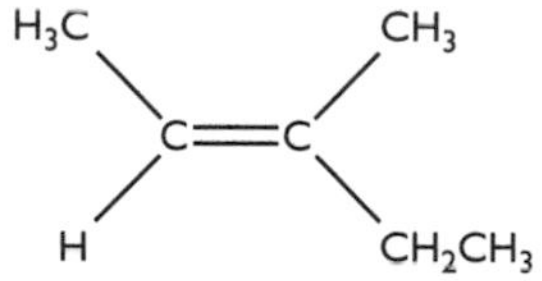

the methyl group on the left-hand carbon atom takes precedence and the ethyl group on the right-hand carbon takes precedence ✓; these are across the double bond, so the isomer is the *E*-isomer ✓ (*entgegen*).

e Note that if this molecule is named using the *cis–trans* notation, it is the *cis*-isomer. The molecule Is (*E*)-3-methylpent-2-ene.

In this question many of the parts depend on the previous part. So, if you think that compound A is but-1-ene, you might forget Markovnikoff's rule and think that compound B is a primary halogenoalkane and that compound C is a primary alcohol, which oxidises to an aldehyde or to a carboxylic acid. This would be taken into account in the mark scheme, so that marks could still be scored, even though an initial error had been made.

Question 14

Explain in terms of the intermolecular forces present the following observations:

(a) Ethanol has a much higher boiling temperature (352 K) than propane (231 K), even though the molecules are of similar size. (4 marks)

(b) Ethanol is less polar than chloroethane, but ethanol is soluble in water whereas chloroethane is not. (4 marks)

Total: 8 marks

■ ■ ■

Answer to question 14

(a) The molecules are similar in size so they have the same number of electrons and the dispersion forces are similar. ✓ Ethanol forms intermolecular hydrogen bonds ✓, whereas propane has dispersion forces only ✓, so more energy is required to separate the ethanol molecules ✓.

ℓ The question is a comparison between two molecules, so both must be referred to in your answer. Weaker answers often ignore one of the molecules.

(b) Ethanol can form hydrogen bonds with water ✓, which compensates energetically for the breaking of intermolecular hydrogen bonds in water itself ✓. Chloroethane cannot form hydrogen bonds with water since chlorine is not sufficiently electronegative ✓, so the (endothermic) disruption of intermolecular hydrogen bonds in water cannot be offset by (exothermic) solvent–solute bonding ✓.

ℓ It is important to recognise that dissolving requires solvent–solvent bonds to be broken and that the energy required to do this has to come from formation of a solvent–solute interaction.

Question 15

(a) (i) **Draw the shape of the ammonia molecule (NH$_3$) and mark on your drawing the approximate value of the H–N–H bond angle.**   (2 marks)

   (ii) **Explain why ammonia has this shape and why the bond angle has the value you state.**   (4 marks)

(b) **Explain the meaning of the term *electronegativity*.**   (1 mark)

(c) **Tetrachloromethane (CCl$_4$) has polar C–Cl bonds.**

   (i) **Explain why C–Cl bonds are polar.**   (1 mark)

   (ii) **Explain whether or not the CCl$_4$ molecule is polar overall.**   (2 marks)

Total: 10 marks

■ ■ ■

Answers to question 15

(a) (i)

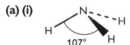

Ammonia molecule drawn pyramidal ✓ and the bond angle marked as being between 105° and 108° ✓.

   (ii) The molecule has three bond pairs and one lone pair ✓ arranged as far apart as possible *or* to the position of minimum repulsion ✓, lone pair–bond pair repulsions are greater than bond pair–bond pair repulsions ✓, so the bond angle is reduced from the tetrahedral angle (of 109.5°) *or* the carbon–hydrogen bonds are pushed closer together ✓.

Valence-shell electron-pair repulsion theory makes clear that the molecular shape depends on the electron pairs, not the bonded atoms, so the pairs must be explicitly described. A common error in weaker answers is the suggestion that the bonds are arranged to give maximum repulsion — do not fall into this trap.

(b) The attraction of a bonded atom for the electron pair in the bond ✓

(c) (i) The electronegativity of chlorine is greater than that of carbon so electrons are unequally shared. ✓

   (ii) The molecule is tetrahedral and therefore symmetrical ✓ so the bond polarities cancel and the molecule is not polar ✓.

Question 16

(a) **State and explain the trend in the first ionisation energy for the elements of group 2, Mg to Ba.** (3 marks)

(b) **The thermal stability of the nitrates of group 2 increases from magnesium nitrate to barium nitrate.**

  (i) **Give the equation for the thermal decomposition of magnesium nitrate.** (2 marks)

  (ii) **Explain why magnesium nitrate is less thermally stable than barium nitrate.** (3 marks)

(c) (i) **State the flame colour produced by barium ions.** (1 mark)

  (ii) **Explain how a flame colour such as that from barium is produced by processes within the atom or ion.** (2 marks)

  (iii) **Suggest an analytical process that depends on flame colour other than the flame test itself.** (1 mark)

Total: 12 marks

■ ■ ■

Answers to question 16

(a) The first ionisation energy becomes less endothermic *or* falls ✓ as the group is descended *or* as the atomic number of the element becomes larger ✓, since the outer electron is further from the nucleus and is increasingly shielded by *or* repelled by the inner electrons ✓.

🄴 A perennial problem with answers concerning thermochemistry (and later in the course in electrochemistry, where voltages can be positive or negative) is what is meant when the values 'fall'. Most would say that $+500$ kJ mol^{-1} is larger than $+200$ kJ mol^{-1}; but is -500 kJ mol^{-1} greater or smaller than -200 kJ mol^{-1}? The sign indicates the *direction* of movement of heat energy, $+$ into the system and $-$ out of it. So -500 kJ mol^{-1} is a loss from the system of a greater amount of heat energy than -200 kJ mol^{-1}. In this case, it is better to say that the former value is *more exothermic* than the latter. So the first answer above is the better one.

(b) (i) $Mg(NO_3)_2 \rightarrow MgO + 2NO_2 + O_2$

🄴 ✓ for MgO and NO$_2$, and ✓ for correct balancing.

Some chemists object to the idea that 'down the group' is equivalent to increasing atomic number, i.e. from Mg to Ba in group 2. It is true that the periodic table does not have to be written in the conventional manner, but only a real pedant would object to the notion that 'down group 2' means anything other than from Mg (or Be) to Ba. However, to be fail-safe, spell it out as in the stem of questions (a) and (b).

State symbols are not needed here as the question does not ask for them. If they are required, a mark would be allocated for giving them correctly. If you are not asked to give state symbols, it is better to leave them out. Although they would be ignored if correct, there might be a problem if they are wrong. The equation with state symbols is:

$$Mg(NO_3)_2(s) \rightarrow MgO(s) + 2NO_2(g) + O_2(g)$$

(ii) The magnesium ion is smaller than the barium ion ✓, so magnesium ions have a higher charge density and so are more polarising (than barium ions) ✓ and they distort the electron distribution in the nitrate ion more effectively (and therefore magnesium nitrate decomposes at a lower temperature) ✓.

☑ It is not enough for the first 2 marks to say that the charge density of a magnesium ion is higher than that of a barium ion.

Be careful to write magnesium *ion* or barium *ion*. The statement that 'magnesium has a higher charge density than barium' is often seen, but never credited.

(c) (i) (Apple) green ✓

☑ Some apples are indeed other colours, but if it is a green apple, that's the colour of the barium flame!

(ii) Heating the atom/ion in the flame promotes electrons to higher energy levels ✓. Electrons falling back down emit their energy as light ✓.

(iii) Analysis of the amount of potassium or sodium ions in biological fluids using a flame photometer ✓

☑ Any other sensible use would be accepted, but not its use in fireworks, since that is not an analytical application.

Question 17

Spectroscopy is a powerful tool in the determination of molecular structure. For organic molecules mass spectroscopy and infrared spectroscopy are widely used.

(a) An organic compound **X** is believed to be either butanone or butan-2-ol. Its mass spectrum is shown below.

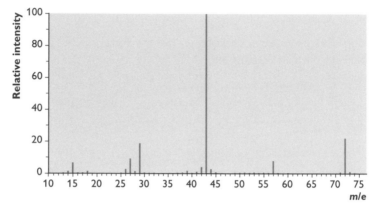

   (i) Explain why, on the basis of the peak at *m/e* 72, the compound is butanone rather than butan-2-ol.   (3 marks)

  (ii) Give the formulae of the fragment ions that are responsible for the peaks at *m/e* 43 and *m/e* 29.   (2 marks)

(b) Infrared spectroscopy can readily distinguish alcohols such as butan-2-ol from ketones such as butanone.

   (i) State what property a bond must possess if it is to absorb in the infrared region of the spectrum.   (1 mark)

  (ii) With the aid of your *Data Booklet* explain whether the spectrum shown below is that of butan-2-ol or of butanone.   (2 marks)

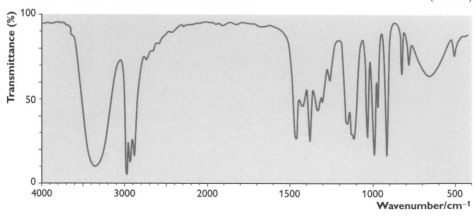

Total: 8 marks

Answers to question 17

(a) (i) The peak at *m/e* 72 is the molecular ion peak ✓ corresponding to the relative molecular mass of butanone ✓, so the ion is $CH_3COCH_2CH_3^+$ ✓.

   (ii) 43 is CH_3CO^+ ✓ and 29 $CH_3CH_2^+$ ✓.

🖉 Note carefully the positive charges on the ions. Weaker candidates often forget to include these charges, costing them a mark.

(b) (i) The bond must change its polarity when it vibrates ✓.

   (ii) The broad peak at 3200–3500 cm⁻¹ ✓ is characteristic of hydrogen-bonded alcohols, so the spectrum is that of butan-2-ol ✓.

Context questions

In past context questions the context was often supplied as a short statement, rather than a longer passage. Note particularly that the passage gives the context but the question is about the chemistry; this is not a comprehension exercise. Extended writing is also required.

Question 18

The halogens (group 7) are important elements that are used widely. Chlorine is the most significant element in the maintenance of public health because of its use in the disinfection of water supplies. In early Victorian times, cholera was common in London and elsewhere and it was through proper sewage disposal and also the treatment of drinking water with chlorine that the disease was eliminated. Chlorine and its compounds are important as bleaches and disinfectants, and in the extraction of bromine. Halogenated compounds are used widely as anti-stick coatings, anaesthetics, flame-proofing agents, photographic chemicals, for the manufacture of window frames and piping and as intermediates in chemical synthesis in the pharmaceutical industry. The 1906 Nobel prize in chemistry was awarded to Henri Moissan for his discovery, in 1886, of fluorine, the most reactive of all elements.

(a) Seawater contains aqueous bromide ions. During the manufacture of bromine, seawater is treated with chlorine gas and the following reaction occurs:

$$2Br^- + Cl_2 \rightarrow Br_2 + 2Cl^-$$

    (i) Explain the term oxidation in terms of electron transfer. (1 mark)

    (ii) Explain the term oxidising agent in terms of electron transfer. (1 mark)

    (iii) State which of the elements, chlorine or bromine, is the stronger oxidising agent.
    Explain the importance of this in the extraction of bromine from seawater, as represented by the equation above. (2 marks)

(b) When sodium chlorate(I), NaClO, is heated, sodium chlorate(V) and sodium chloride are formed.

    (i) Write the ionic equation for this reaction. (2 marks)

    (ii) What type of reaction is this? (1 mark)

(c) During the process for the manufacture of iodine the following reaction occurs:

$$2IO_3^- + 5SO_2 + 4H_2O \rightarrow I_2 + 8H^+ + 5SO_4^{2-}$$

    (i) State the oxidation numbers of sulfur in SO_2 and SO_4^{2-}. (2 marks)

    (ii) Use your answer to part (c)(i) to explain whether SO_2 is oxidised or reduced in the above reaction. (2 marks)

    (iii) Name a reagent that could be used to confirm that a solution contains iodine. State what you would see. (2 marks)

(d) Chlorofluorocarbons, or CFCs, have been used widely as refrigerants and in aerosol propellants. Their use is appealing because they are inert and non-toxic. However, their use is being discontinued because they are environmentally damaging. Explain the circumstances under which CFCs harm the environment. Illustrate your answer with suitable equations. (5 marks)

Total: 18 marks

■ ■ ■

Answers to question 18

(a) (i) Oxidation is electron loss ✓.
  (ii) An oxidising agent gains electrons ✓.
  (iii) Chlorine is the stronger oxidising agent ✓; it removes electrons from bromide ions ✓ to produce bromine.

🖉 It is important to distinguish between bromine and bromide. Always make clear which species is under consideration. Similarly 'magnesium' is an element — if you mean 'magnesium ions' (Mg^{2+}), you must include the word 'ions'.

(b) (i) $3ClO^- \rightarrow 2Cl^- + ClO_3^-$

🖉 There is 1 mark for the species and 1 mark for the balanced equation. Note that the question asks specifically for the ionic equation. The non-ionic version including the sodium might get 1 mark.

  (ii) Disproportionation ✓

🖉 It is true that this is also a redox reaction. However, it is a particular type of redox reaction and this is the more significant piece of information.

(c) (i) In SO_2 the oxidation number of sulfur is +4 ✓; in SO_4^{2-} it is +6 ✓.

🖉 Note that there is no credit for showing any working because there is only 1 mark available for each answer. This is unlike other calculation-type questions in which there are often marks available for working, even if the answer is wrong.

  (ii) Oxidised ✓ because the oxidation state of sulfur has increased ✓.
  (iii) Starch ✓; it turns blue-black ✓.

🖉 Starch-iodide paper should not be used in this test; the iodide ions could be oxidised to iodine by either chlorine or bromine, so starch-iodide paper turns blue-black with all the halogens. Starch turns blue-black only with iodine.

(d) —

🖉 There are many answers to this question. Your answer should have six of the following scoring points (✓) written in a logical manner and in good English; at least one of the scoring points must be an equation. Quality of written communication is always assessed in the context question.

- CFCs are decomposed by ultraviolet light ✓ to give chlorine radicals.
- $CF_3Cl \rightarrow {}^{\bullet}CF_3 + Cl^{\bullet}$ ✓
- Chlorine radicals catalyse the decomposition of ozone ✓.
- One chlorine radical can destroy many ozone molecules ✓.
- $Cl^{\bullet} + O_3 \rightarrow ClO^{\bullet} + O_2$ ✓
- $ClO^{\bullet} + O^{\bullet} \rightarrow Cl^{\bullet} + O_2$ ✓
- Ozone protects the surface of the Earth against harmful ultraviolet radiation ✓.
- Higher ultraviolet levels could increase the incidence of skin cancer ✓.
- CFC molecules absorb infrared radiation ✓ because their polarity alters when the bonds vibrate ✓.
- CFCs are, therefore, greenhouse gases ✓.

When answering questions about environmental matters such as the greenhouse effect, try to retain a sense of proportion. Many student answers are almost apocalyptic. This is not appropriate in a science examination. Remember that the science behind ozone depletion and the absorption of infrared radiation may be good, but the predictions of the long-term effects are controversial.

Question 19

Ethanal (CH_3CHO) is an important organic substance. It is used in the manufacture of perfumes, dyes, polymers, paints and pharmaceuticals and many other substances. Much of it is oxidised to ethanoic acid for use in food, and it is one of the materials used in the process for silvering mirrors. It occurs naturally by oxidation of ethanol in wine, as well as in ripening fruit, coffee, grapefruit, car exhausts and in many essential oils, including those of rosemary, fennel and peppermint. Ethanal can be made in a number of ways, two of which are given below.

(1) **By oxidation of ethanol vapour over a heated copper catalyst:**

$$CH_3CH_2OH \rightarrow CH_3CHO + H_2$$

(2) **By the oxidation of ethanol using potassium dichromate(VI) in dilute sulfuric acid, distilling off the ethanal as it is formed. The dichromate(VI) ions are reduced to chromium(III) ions (Cr^{3+}).**

$$Cr_2O_7^{2-} + 3CH_3CH_2OH + 8H^+ \rightarrow 2Cr^{3+} + 3CH_3CHO + 7H_2O$$

The chromium(III) compound produced has little commercial value compared with ethanal, and dichromate ions are toxic. The reaction is suitable for laboratory use, the ethanal being distilled off from the reaction mixture as it is formed by addition of potassium dichromate(VI) and ethanol to hot sulfuric acid.

(a) Suggest three reasons why reaction (2) is a less desirable method for making ethanal on an industrial scale than reaction (1). (3 marks)

(b) The ability to distil out ethanal before it oxidises further depends on its having a lower boiling temperature than ethanol. Explain in terms of the types of intermolecular forces present why an aldehyde always has a lower boiling temperature than the alcohol from which it can be made. (3 marks)

(c) Ethanol can also be heated under reflux with potassium dichromate(VI) and dilute sulfuric acid.

    (i) Explain what *heating under reflux* is and why it is used in preparative organic chemistry. (3 marks)

    (ii) Give the full structural formula of the organic compound obtained by heating ethanol under reflux with potassium dichromate(VI) and dilute sulfuric acid. (2 marks)

    (iii) Give the full structural formula of the secondary alcohol having three carbon atoms. (1 mark)

    (iv) Give the full structural formula of the organic product from heating the alcohol in (iii) under reflux with potassium dichromate(VI) and dilute sulfuric acid. (1 mark)

Total: 13 marks

■ ■ ■

Answers to question 19

(a) *Three of the following:*

- The chromium(III) compound is a waste product in reaction (2) since it has no major use. ✓
- It is expensive to safely dispose of the chromium(III) compound from reaction (2). ✓
- Reaction (2) has a poor atom economy, since a significant number of the atoms used at the beginning do not finish up in the product compared with reaction (1). ✓
- Both of the products from reaction (1) are useful materials. ✓
- The presence of dichromate(VI) ions in effluent would be a serious health hazard. ✓

🖉 There are other possible scoring points. The acceptable form of words and scoring points would be decided by the senior examining team at its meeting. In their preliminary marking of scripts examiners might come across ideas that had not been considered when the paper was set.

(b) The alcohol has intermolecular hydrogen bonds ✓, whereas the aldehyde has dipole–dipole forces ✓. Hydrogen bonding is stronger than dipole–dipole so more energy is needed to overcome it. ✓

🖉 Note that when you answer questions concerning two compounds, you need to ensure that you mention both of them.

(c) (i) The reaction flask is fitted with a vertical condenser ✓, which returns the condensed liquid continuously to the flask ✓. This enables prolonged heating without loss of solvent. ✓

(ii)

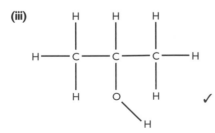

🖉 ✓ for giving a carboxylic acid and ✓ for giving ethanoic acid.

(iii)

The bond in the O–H group is shown because a full structural formula is required. Showing the group as –OH would probably be accepted, but it is best not to test this idea.

(iv)

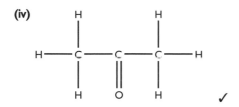

✓